THE SENTIENT MACHINE

THE COMING AGE OF
ARTIFICIAL INTELLIGENCE

AMIR HUSAIN

SCRIBNER

New York   London   Toronto   Sydney   New Delhi

Scribner
An Imprint of Simon & Schuster, Inc.
1230 Avenue of the Americas
New York, NY 10020

First Scribner hardcover edition November 2017

SCRIBNER and design are registered trademarks of The Gale Group, Inc.,
used under license by Simon & Schuster, Inc., the publisher of this work.

For information about special discounts for bulk purchases,
please contact Simon & Schuster Special Sales at 1-866-506-1949
or business@simonandschuster.com.

The Simon & Schuster Speakers Bureau can bring authors to your live event.
For more information or to book an event, contact the Simon & Schuster Speakers
Bureau at 1-866-248-3049 or visit our website at www.simonspeakers.com.

Manufactured in the United States of America

1   3   5   7   9   10   8   6   4   2

Library of Congress Cataloging-in-Publication Data has been applied for.

ISBN 978-1-5011-4467-7
ISBN 978-1-5011-4469-1 (ebook)

Dedicated to
the children of tomorrow.
May you elevate
the human condition.

CONTENTS

PROLOGUE:
A BOY'S DISCOVERY

I remember exactly where I was when I first saw *it*. I was four years old, visiting my friend's house in Lahore, Pakistan, and the glimmer of its flickering screen caught my eye. There it was, sitting on their TV console: the Commodore 64—the most popular personal computer on the market in 1982. It was connected to a television screen and running the game Hangman, an early version of the video games we know so well today. The television was a fixture in my home and I knew that it played images and sounds on the screen. But this machine was worlds away from mere TV. Television was, in a sense, immutable. Predetermined. This machine, on the other hand, was acting as a result of my inputs. This was a device I could impact with my own ideas. Ideas could flow through my fingertips and end up on that screen.

As soon as we left my friend's house, I returned home and immediately set to work with spare pieces of broken and discarded toys and old cardboard boxes and packages. "Look," I called out to my parents, "I made a computer!" The discovery of this machine felt as tactile as holding a paintbrush or a piece of modeling clay in my hand. A computer could serve as my ultimate tool of creativity. This was the means by which I would impact the world.

Since that day in 1982, I've never wanted to do anything else. I never wanted to be a fireman or a doctor or an astronaut. Today, even after having lived thirty-eight years immersing myself in this one pursuit, I am nowhere close to being done. Computing is one of the great drivers in my life.

Some of you will immediately relate to my experience with the Commodore 64, while others will find this whole tale quite foreign. For those of you who don't have an affinity for computers and the science around them, I would like to invite you into my world. Before we can begin to discuss the future of artificial intelligence and our role, as humans, in relation to these machines, we should first take a moment to appreciate what makes computing concepts so elegant and wondrous. My goal is to communicate the beauty inherent in this way of thinking about the world so that you might better appreciate how humanity can live and thrive amid this coming age of sentient machines.

One of my influences in the computer science department at the University of Texas at Austin, and one of the greatest computer scientists in history, Edsger Dijkstra, argued that we were not studying computer science but rather *computing* science. In this way, he acknowledged that the computer was an outgrowth of a perspective on the world, a way of contending with reality. His approach took computer science outside the realm of technical understanding and placed it firmly in the grip of some of the deepest and most profound concepts in our understanding of existence.

Needless to say, computer science is not just about devices. This became startlingly clear to me when I was around eleven years old. My father handed me a general-interest magazine called *Dialogue,* distributed in Pakistan by the United States Information Service (USIS). In the pre-Internet days, books and publications like these were some of the ways to get information, so though the magazine itself was nothing extraordinary, the information inside was precious. I opened it up and immediately took in the title of one of the main features about computer scientist and physicist Ed Fredkin: "Is the Universe a Computer?" In those five words, a notion with profound explanatory power came together for me. It confirmed that the richest concepts in computer science come to us directly from nature.

Take programmability, for example. When you want to build something, you have two options: the first is to simply build it yourself by directly carrying out the necessary steps; the second is to create a machine that can carry out the steps for you. When you write a program, you are essentially providing a recipe to something that can interpret and carry out many types of these recipes, or repeatedly apply the same one. Change a few words here or a command there, and the output can become entirely different. This ability to transform outcomes with ease is the essence of programmability. The instructions processed by a computer are the programs; inside these programs are codified ideas that solve problems like sorting numbers, searching text, or transforming images. These are known as algorithms. With systems like computers, there is the flexibility of using programming to build not just one thing but many, many things. What if a program could write itself? What if the system interpreting a program—also a program—could be modified? In this universe, everything is fungible with little to no cost. A programmer can simply will things into being.

This very same concept of programmability is also reflected

in biology and the natural world. Think of DNA: the ultimate code. In fact, all complex biological forms in nature are consequences of computations and transformations directed by DNA. Fractals are another example, as these never-ending patterns are self-similar across all different scales. There is no way for a human to pick up a paintbrush and paint a "complete" fractal. You can only ever employ computation to create its form. Of course, nature creates shapes like this all the time in snowflakes, seashells, clouds, trees, and coastlines. So is nature, then, a computer?

When I first read "Is the Universe a Computer?" it set me on a quest toward understanding how to build a universe. Just like a fractal, you can't build a universe by specifying every element of it directly. You can only build something this complex by specifying the processes and then allowing the incredibly powerful concepts of iteration and recursion to take over.

Consider a famous example called The Game of Life. For those of you deeply immersed in the world of computer programming, mathematician John Conway's iconic game of cellular automata will be very familiar. The Game of Life, or Life, as it was often called, contained an infinite grid with rows of cells—first printed in static form in the pages of *Scientific American* in 1970 and then later run through computers—with four or five simple rule sets. Each cell can be in a state of "alive" or "dead" and each cell interacts with its eight neighbors: the cells that are horizontally, vertically, and diagonally adjacent to it:

1. Any live cell with fewer than two live neighbors dies, as if caused by underpopulation.
2. Any live cell with two or three live neighbors lives on to the next generation.

3. Any live cell with more than three live neighbors dies, as if by overpopulation.

4. Any dead cell with exactly three live neighbors becomes a live cell, as if by reproduction.

The Game of Life is just one example of cellular automata. There are many more, such as those chronicled by Stephen Wolfram in his book *A New Kind of Science*. With many of these cellular automata, often just six to eight simple rules govern how cells turn on or off. But the rules also generate things that appear to be patterns. Visually, these are not random, not noise. An identifiable picture appears and it is never-ending: the metapattern keeps repeating but the specific details of each and every pattern never repeat. It is infinite newness with the minimal amount of input. This is the iterative application of a simple rule.

Computers have even generated Mandelbrot fractals, named after the French-American mathematician Benoit Mandelbrot, that exceed the size of the known universe. Imagine that: you can spend your entire life just traversing the edges of a computer-generated construction—akin to an ancient traveler along the fabled coastlines described in a Greek myth. At the end of your lifetime, you still would not have seen it all. The secrets of these forms are unending, a fact that fills me with wonder, humility, and awe. And when I see them applied in practice, I realize that core computer science concepts—recursion, iteration, abstraction, generative programming, and many others I will discuss throughout this book—are the true wealth of our humanity at its most creative. They give us a rich bounty through which to understand the furthest corners of both our world and our mind, and they form the intellectual scaffolding that runs throughout this book. We can use these concepts to solve some of the greatest challenges of our world today and tomorrow. Most important, humanity can use computational science to achieve our ultimate purpose: to explore, to create, and to understand our universe.

PART ONE

What Is AI?

The first time it happened, I was at the University of Texas in Austin as a student. It was one of my first weeks of school and I woke up in my bedroom in such a confusing paroxysm of pain that I was convinced I was dying. Nothing in my life leading up to that moment had prepared me for this pain: as if a knife were slicing through my right eye socket.

Somehow I stumbled out onto the street where I was able to find my way to the university health center. They handed me an Advil and directed me to lie down in a dark, quiet room until the pain subsided. But the pain did not subside. For two long days, I played mind games with myself in that dark room. *Would you rather* . . . Would you rather stick your hand in an open flame or feel this particular pain in the right side of your head? Would you rather have a bullet go through your body or continue to experience this throbbing in your eye? These were not abstract questions. They were very real negotiations as I navigated the incomprehensible level of suffering. Eventually, when the throbbing continued unabated, the game simplified: Would you rather be enduring this torture or be dead?

There is a reason my ultimate diagnosis—I was suffering from the first of many cluster headaches—is referred to colloquially as a "suicide headache." Many headache doctors who have treated patients with chronic cluster headaches have had at least a few of their patients choose death over the pain.

After my first episode with them at the age of eighteen, I didn't have them again until I was twenty. Both of those attacks were relatively short—lasting only a few days. Over time, however, the duration of the attacks increased. They went from days to months. I couldn't find words or numbers to define this type of pain. Simply put, it was impossible to even remain conscious through it. Eventually, I almost always passed out.

Two years ago, after a remission of several years, I was revisited by my old demons. One night, I experienced an attack so extreme that I knew I was in trouble. I had gone way beyond my weekly dosage of the prescribed sumatriptan, so I reached for another painkiller in desperation. My vision got cloudy; the last thing I remember is telling my wife to call 911.

When I woke up, I was checked into the hospital and the neurologist on staff, Dr. Reddiah Mummaneni, was sitting with me. I could immediately sense that he was kind and bright and engaged. Dr. Mummaneni had done some research on my condition and he handed me a medical paper he had found using a query on his research repository.

"There is only one thing I've found that might help right now," he told me. "It's a really drastic treatment and only one or two incidents are described in this paper. We can try a really high dose of a steroid, methylprednisolone, administered through an IV. It's a common treatment for multiple sclerosis because it acts like a shock to the immune system to calm it down and it helps with the inflammation." He explained that he hoped the anti-inflammatory capabilities of the steroid—at such an extremely high level—could suppress the unexplained inflammation and give a jolt to the mechanisms in my brain that were causing the clusters. After "jolting" my brain for three days, we would move down to an oral steroid for a month and then taper off from there. Dr. Mummaneni said, "I want you to know that this treatment could have very serious repercussions. Are you okay with the risks?" I read the paper and signed off immediately. Considering my current quality of life, there was not a single doubt in my mind. The course of treatment was so severe, however, that the staff was repeatedly instructed to check my sugar, blood pressure, pulse, and heart rate. They followed up with an EKG before, at last, the IV drip was put in place. Suddenly, as quickly as it had started, the attack abated. After four months, the pain was gone. I was finally free. For the time being, at least.

I am a computer scientist, a technologist, and an inventor. I hold two dozen awarded patents and have dozens more pending. I love machines and I have a deep affinity for how they work and what they can do. Although the story I just shared certainly makes a case for compassionate medical care, more than anything, it illustrates to me our own limitations in the field of healthcare. When humans succeed at sorting through all the data involved in an obscure medical treatment, it is often the result of happenstance: I *happened* to get the right doctor assigned to me on that night in the hospital and that doctor *happened* to get the right combination of words for the query to pull up just the right paper. And then he *happened* to have the time to read it and process it. When I use the word *happenstance*, I don't discredit all the wisdom, discernment, and training that go into being a highly skilled—and in this case, an incredibly effective—medical practitioner. *Happenstance* is simply a reflection of the probabilities in search queries that now exist for anyone in the medical field. Medline, one of the most popular online research repositories for doctors, reported that more than 6 million journal article references have accumulated since 1965 and their database is growing at a rate of 300,000 per year. Human intelligence is simply too limited to process this amount of information. If we happen to find the essential data point, we find it. If we happen to miss it, we miss it.

But there is a different kind of intelligence much better suited to solving these types of challenges; this is artificial intelligence. Today, AI is being applied to all kinds of health data, allowing us to assess previously invisible, seemingly immeasurable aspects of our lives. Why did my cluster headaches go into remission for two years and then one year and then five years? Why did they sometimes occur at the same time twice a day and sometimes only once a day? Why did they strike in the fall sometimes and at other times in the summer? What is really happening? Can artificial intelligence help me solve my own personal mysteries? Can I use my intimate knowledge of computer science and technology to end my suffering?

Cluster headaches are just one of countless ailments that cause unbearable pain but don't warrant major research or attention because not enough people suffer from them. Our human intellect is a finite resource and it is currently—and rightly—directed at problems that have the biggest impact on humanity. But there are lots of diseases that will never receive the top-end human capital and expertise because they simply don't affect enough people. For that reason, many people who experience the long tail of suffering will never receive the quality of intellect they need to get rid of their pain. Humans simply can't keep up with all the diseases. But machines can. And they must—if we want to improve our lives.

This isn't just happening in healthcare. This is happening in every domain in our society. This is happening in astronomy, chemistry, materials science, manufacturing, financial services. The list goes on and on. This is *progress*, and I believe it is imperative we embrace it.

WHAT ARE WE SO AFRAID OF?

I was speaking on artificial intelligence at the popular technology conference South by Southwest—SXSW—in my hometown of Austin. Soon after, protesters chanted "I say robot; you say no-bot," and their signs read "Stop the Robots!" and "Humans Are the Future!" The year was 2015 but it might as well have been 1980, 1967, 1950, or even the late 1800s. Throughout all of these periods in history, technological advances have left us both optimistic and also deeply uneasy about the role of machines in our culture.

These ebbs and flows of our greater historical mood toward machine automation can be tracked to the technological advancements of the age. More than two centuries ago, the invention of the steam engine sent society into a tailspin as cultural commentators, prognosticators, and businesses tried to parse out the "machinery question." Writers associated with the Young England movement aligned themselves with laborers in sentimentalizing the work of the hand as a patriotic toil. Machines like the steam engine and the power loom became emblematic as dehumanizing forces that threatened the national character.

Merrily went the click-clack, the hammer, and the plough
And honest men could live by the sweat of their brow.

In light of Brexit, this yearning for "merry old England" feels particularly resonant. The destabilizing forces of technological innovations inevitably bring about our nostalgic longings for a simpler time. It was in the midst of this very shift—the Industrial Revolution—that Mary Shelley published her iconic novel of horror in 1818, a work that explored the dehumanizing forces of an increasingly technological society through the lens of literature's first modern scientist: Victor Frankenstein.

Mary Shelley was only eighteen when she wrote *Frankenstein*. In one of the coldest and dreariest summers ever on record—a huge volcano erupted in Indonesia in 1815, sending a vast cloud of ashes across the Northern Hemisphere the following year—Shelley was holed up in an elegant Swiss villa with a bohemian coterie that included Lord Byron, her stepsister Claire Clairmont, and her romantic partner, Percy Shelley. After countless nights of rain and thunder, Lord Byron suggested they should all write ghost stories. Shelley had only recently left the home of her father, the famous progressive William Godwin, where she was exposed to the work of the great scientists of the day, taking in a wide range of ideas in subjects ranging from anatomy to galvanism—the study of muscle contractions as a result of electrical currents. At Byron's suggestion, Shelley's influences and inspirations came together to create her characterization of Victor Frankenstein.

In twentieth-century films and stage plays, Frankenstein's monster was rendered in more melodramatic fashion with pulsing electrodes and nodules on his head. In Shelley's original version, however, scientist Victor Frankenstein uses elemental principles of life to imbue vitality into inanimate matter. She even makes a distinct point of stating that Frankenstein draws upon the scientific method for his work. The moment she describes her vision of the monster, however, is a passage that harkens back to writing in the ancient Greek myth of Prometheus or from Genesis in the Old Testament:

> *I saw the pale student of unhallowed arts kneeling beside the thing he had put together. I saw the hideous phantasm of a man stretched out, and then, on the working of some powerful engine, show signs of life and stir with an uneasy, half-vital motion. Frightful must it be, for supremely frightful would be the effect of any human endeavour to mock the stupendous mechanism of the Creator of the world.*

In this passage, we can see the crux of a key philosophical quandary, the cause of our unease: Who can be a creator? And when does our creation truly acquire agency?

Frankenstein posed these questions to an increasingly modern society, but for as long as humanity has been creating, we have also been in conflict with our role as creators. Many of the world's greatest religions have decrees and practices around aniconism, banning material representations of living creatures and the divine. There are countless passages in the Hebrew Bible against graven imagery, beginning with the Second Commandment:

> *Thou shalt have no other gods before me. Thou shalt not make unto thee any graven image, or any likeness of any thing that is*

in heaven above, or that is in the earth beneath, or that is in the
water under the earth: Thou shalt not bow down thyself to them,
nor serve them: for I the Lord thy God am a jealous God, visiting
the iniquity of the fathers upon the children unto the third and
fourth generation of them that hate me.

And yet Hellenic stories abound with tales of automated representational machinery. In ancient Greek myths, inventions by characters like Hephaestus, the blacksmith of Olympus, are celebrated as wondrous innovations. He even designed what is surely the world's first notion of a fully automated waitstaff with twenty wheeled devices propelling themselves in and out of the halls of Olympus, serving the gods while they indulged in drink and wine.

Historian and writer Pamela McCorduck, who has spent much of her career documenting the history of artificial intelligence, characterizes these two perspectives toward our creation of lifelike machinery—wondrous and useful versus arrogant and dangerous—as the polarities that inform our conflict even to this day. Do we dare usurp that power from a god or the natural world? And if we do, are we prepared to face the unintended consequences?

The Genesis story in the Quran, for example, quite similar to that in the Old Testament, narrates that Adam was cast from clay when God breathed his spirit into the clay form. God then taught Adam the nature of all things. With knowledge and the power to think thus imparted, God deemed Adam the most superior among all creation. According to this particular creation story, God ordered the angels and the fire-being Iblis, or Lucifer, to bow before this new, autonomous creation. When the angels innocently asked why, God explained that it was because Adam could acquire knowledge. His thought was not static, and that allowed him to learn, grow, and change. And so, in this seminal narrative of creation, the first form of autonomous intelligence distinct from God—human intelligence—was unleashed upon the universe.

Today, of course, we deal not with myth and allegory but with the real-world implications of an explosion of artificial intelligence that is outpacing our ability to understand its consequences. Even Silicon Valley's most powerful players seem divided about which side of the technological divide they are on. In 2014, Tesla and SpaceX founder Elon Musk spoke at Massachusetts Institute of Technology and called AI humanity's "biggest existential threat." His talk went on to accuse artificial intelligence of "summoning the demon."

"You know all those stories where there's the guy with the pentagram and the holy water and he's like, yeah, he's sure he can control the demon?" he asked the crowd before saying, "Doesn't work out."

Elon Musk is not alone in sounding out the alarms. He is joined by intellectual heavyweights like Stephen Hawking, Bill Gates, Oxford professor and existential philosopher Nick Bostrom, and Henry Kissinger. In 2014, Hawking told the BBC: "The development of full artificial intelligence could spell the end of the human race," and in a 2015 interview session on the popular Silicon Valley website Reddit, Bill Gates added, "I am in the camp that is concerned about super intelligence. First the machines will do a lot of jobs for us and not be super-intelligent. That should be positive if we manage it well. A few decades after that, though, the intelligence is strong enough to be a concern. I agree with Elon Musk and some others on this and don't understand why some people are not concerned."

In late 2015, Musk announced his intent to provide $1 billion in funding to a new nonprofit dedicated to safe artificial intelligence research. The organization, called OpenAI, brings together talented practitioners in the field and proposes to make their designs and code publicly available. Soon after, Musk, Hawking, and a group of a thousand other tech titans and power players signed a letter calling for a ban on autonomous weapons. In the fall of 2015, Silicon Valley's most prominent technology companies came together to create the Partnership on Artificial Intelligence, exploring all of the issues, dangers, and ethical concerns that surround this coming age of intelligent machinery.

In our journey together, we will explore how anti-AI movements could be a threat to developing much-needed technology to solve this century's most complex problems. In Part Two, I will explain why trying to suppress such work, or subject it to draconian regulation, will be incredibly harmful to us as a civilization. Before we approach these coming challenges, however, we need to situate ourselves in the culture of fear that surrounds our notions of artificial intelligence. And we need to fully understand how algorithms excel at tasks that human brains find so daunting. What is machine intelligence exactly?

Too often, we frame our discussions of AI around its anthropocentric characteristics: How much does it resemble us? Can it "pass" as human? This kind of thinking can be traced back to the 1950s when the preeminent mathematician Alan Turing, memorialized in the recent film *The Imitation Game*, published his paper "Computing Machinery and Intelligence" and described what we now call the Turing Test. In Turing's test, a human interrogator interacts with both a human and a machine using a text-only channel, such as a screen. The computer succeeds—according to Turing's paper—when it can fool the human interrogator with its answers, making the interrogator believe that it is human. Versions of the Turing Test as a barometer of success still abound all around us. In voice-recognition software programs, it's considered a major coup when humans speak into the phone, imagining that they are speaking to a friendly sales representative in Nebraska, not a computer algorithm.

This is a curious form of narcissism. Do we really imagine that human intelligence is the only kind of intelligence worth imitating? Is mimicry really the ultimate goal? Machines have much to teach us about "thoughts" that have nothing to do with human thought.

The Turing Test and other such "mimicry" metrics for machine learning may be less relevant in today's practical applications of artificial intelligence. But they still keep our cultural discourse captive—just think of all the recent films that involve a human "falling" for the powerful mimicry of a machine. Steven Spielberg's *A.I.* and Spike Jonze's *Her* are only two of many examples.

We need to take a moment to explore how "thoughts" can be engineered in an entirely different way with artificial intelligence. We may think we understand "cognitive diversity" in a human realm, and yet different human thinkers all have an amygdala, which stimulates fear; a prefrontal cortex, responsible for simulating future happenings; and a visual cortex that transforms the data from our eyes into images. Machines work under a very different model—even their silicon substrate is different from the carbon substrate that makes up our own brains. As a result, the intrinsic strengths of the machine-mind—speed, unconstrained energy consumption, limitless recall (short-term and long-term memory)—are inherently different from those of the biological mind. Where are we in AI today? And how far away are we from a genuinely intelligent form of AI?

ARTIFICIAL INTELLIGENCE, MACHINE LEARNING, AND COGNITIVE COMPUTING: HOW DOES HUMAN INTELLIGENCE COMPARE?

There has been a flood of interest in artificial intelligence recently and the popular press uses terms such as cognitive computing and machine learning almost as synonyms for AI. It is useful to understand the true meaning of these terms, as they are not merely many names for the same thing. In fact, understanding the differences between these terms aids in our ability to differentiate machine-based intelligence from human thought.

Artificial Intelligence is a broad field of study, formally "founded" by John McCarthy and Marvin Minsky following a now famous 1956 Dartmouth summer conference involving such topics as mathematics, game theory, and logic. The field has traditionally studied many different types of techniques that can be implemented in machines, enabling them to reason, learn, and act intelligently. Some of these techniques can act based on knowledge and rule sets represented at the time of the construction of a system. Others can use "heuristics"—often, informed guesses—to search through a large set of potential choices, selecting the one that appears to be most reasonable. And still other approaches start off with some core assumptions and then factor in new information—they "learn" as they go.

Artificial Intelligence is the overarching science that is concerned with intelligent algorithms, whether or not they learn from data. In contrast, **Machine Learning** is a subfield of AI devoted to algorithms that learn from data. This learning can fall into multiple categories. For example, in "supervised" learning, an algorithm is presented with a set of prior, labeled examples and has the benefit of identifying associations between the data and the labeled outcome, or classification. In "unsupervised" learning, prior sets of labeled examples are not available, but unlabeled, or uncategorized, data are. Think of supervised learning as having the goal of differentiating future pictures of cats and dogs after first being supplied with many pictures of cats and dogs, each labeled "cat" or "dog." Unsupervised learning, by contrast, has the goal of highlighting the differences between a large collection of cat and dog pictures, with no such labels.

Machine learning techniques are concerned with using features or attributes from each example to arrive at the correct label or classification—for example, which key features can be used to distinguish a picture of a cat from a dog? As more examples of cats and dogs are provided, machine learning algorithms can attempt to build models of what underlying distinguishing elements—features—are reliable predictors of whether something is a cat or a dog.

Now, in contrast, consider algorithms that still act "intelligently" but do not use data to learn. One such example would be a chess-playing algorithm that is provided the rules of chess and some way of categorizing whether the board is in a favorable position or an unfavorable position. In the event that there is no precise, scientific way to exactly ascertain the degree to which a board state is favorable, "heuristics," or gut rules, can be used. For example, it is generally preferable to have more pieces remaining, compared to those of the opponent. Similarly, it is generally preferable to have your queen still alive. And it is preferable not to have your king under mate. A chess-playing algorithm could use the rules of chess to construct all possible states that the board could take given the current position, and then assign a rating to each state—whether it is favorable or not. It could then choose to make the move that appears *most* favorable, and repeat until victory or defeat. This is highly simplistic because, among many other things, looking ahead one move is not sufficient—you might want to calculate ahead by several moves. But in general, this is an example of how an algorithm can exhibit seemingly intelligent behavior, and yet not learn from outcomes or new information.

Cognitive Computing, finally, is a term that has been used academically since at least the 1980s, but has become common parlance in the industry as the moniker under which IBM has popularized its Watson system. It is best defined as an area of work that seeks to apply machine learning techniques that are inspired by the human brain. As such, it has a hardware and software dimension. In fact, key projects at IBM that fall under the cognitive computing effort are the TrueNorth architecture, a new type of processor that attempts to implement brain-like structures in silicon, and the more widely known Watson software.

Through the examples above, it is clear that these forms of intelligent algorithms behave quite differently from the human mind. For example, our hypothetical chess-playing algorithm attempts to create all possible board states, or at least a very large subset, given a single, current state and its knowledge of the rules of chess. The human brain doesn't do this. We tend to "prune" possibilities radically because some possibilities seem entirely silly, or not worth much consideration. Algorithms running on very fast computers can afford to explore a larger space of possibilities than the human brain. Our brains, though highly parallelized and excellent at a variety of cognitive tasks, run at a slower "clock speed" than a silicon-based processor, making them slower and less accurate at purely mathematical tasks. In certain cases, this difference between brains and computers—that is, the ability to explore "all options"—is a huge advantage.

In their 1982 paper published in *AI Magazine*, Douglas B. Lenat, William Sutherland, and James Gibbons, researchers working at Stanford University, showed how this ability to generate and analyze a large set of possibilities led to the discovery of new kinds of three-dimensional microelectronic devices. These discoveries were not previously known to human researchers. The algorithm, called Eurisko, generated them on its own.

This ability to generate and evaluate possibilities is best exploited by a field of study involving genetic algorithms that was pioneered by Stanford professor John Koza. In a 2006 article in *Popular Science*, Koza's work was shared with a mass audience under the provocative—but quite accurate—title: "John Koza Has Built an Invention Machine: Its creations earn patents, outperform humans, and will soon fly to space." With the energy and mathematical resources to explore "all options," Koza's genetic algorithms have designed many circuits, and in many cases have even reverse-engineered patented human inventions, implementing them in fundamentally different ways to avoid the original patents.

Of course, human intelligence has its own considerable strengths. The fundamental human limitation of having a relatively small brain, constrained by the energy and physical space available to it, has resulted in a lot of evolutionary "cleverness." Human intelligence makes use of very effective pruning techniques that prevent the brain from running in overdrive, attempting to process billions of potential situations of which only one or two might be useful. The expense of this brute computational approach is simply not suited to biological beings. However, what is clear is that the availability of practically infinite mathematical ability, and abundant calculating speed, do result in discoveries that our biological brains have not yet made. We are "blindsided" to some degree by the biological imperative to conserve energy and make the best use of our brains, given their physical limitations. In this way, machine thought, which can leverage incredible amounts of energy and space, is fundamentally different.

Just as biology constrains our computational abilities, it also places limitations on our memory, endowing us with limited recall and "lossy" storage. As part of psychological or intelligence tests—or just as a fun challenge—many of us have been given an image of a room to look at for a few seconds, and then asked to recall what the color of the curtains was. Or how many paintings were hanging on the wall. It's hard to remember because remembering every last detail has not been critical to our evolutionary survival. However, computers can remember every last detail of every single image, sound, or fact they have ever been exposed to. Humans learn and forget. Computers can learn and don't have to forget. The notion of "what is important" holds very different meanings in the context of a machine. We tend to remember more of what was important at the time, but machines can remember everything and then determine what part of this exactly preserved experience ends up being important at some later stage. It is not hard to see how actions

24

that result from an intelligence with this total recall can be so different from our own.

Most of us will agree that the ability to learn is an intelligent behavior. Yet learning doesn't have the same connotations for humans and machines. Consider, for example, that we can only learn what we perceive. The input has to arrive through one of our senses for us to even register it; we see or read something, feel it with our fingers, smell or taste it, and so on. We try to extend this ability to directly perceive through the abstract tools of mathematics and logic. However, why is it that so few of us are amazingly brilliant mathematicians or physicists? One reason is that the more abstract our experiences get, the harder it is for us to work with them. We cannot "see" four, five, or ten dimensions. And so very few of us can reason over the domain of four or higher dimensions purely through the leverage of tools such as mathematics. In contrast, all of us can see three dimensions directly and therefore do quite well in navigating this directly perceived space. Once again, machine intelligence is—and will be—quite different in this regard. To put it somewhat euphemistically, for a mind made of math, the ability to apply math is not a limitation. High dimensions can be "perceived," processed, and understood in the same way that three dimensions are. There is no limit to the type of sensors that can bring forth inputs that are accumulated and learned over time. There is also no real limit on the number of such sensors/inputs. Would we be different creatures if we also had eyes on the back of our heads? Probably. Machine intelligence takes these differences to an extreme degree.

Perhaps most essentially, human intelligence is a consequence of our mind. Our mind exists in one location and is firmly affixed to our body. There is a oneness to body and mind in the human form. Machine intelligence, on the other hand, can be entirely disembodied. At a basic level, this means that the intelligence is divorced from the need to protect a body. But beyond this, the intelligence can also be copied or be present at multiple locations at the same time. We do not "know" the experience of being present at eleven locations all at once. Machine intelligence will.

Self-improvement is another point of differentiation between human and machine intelligence. Humans have strived for it over millennia. We respect our scholars, teachers, and guides because they help us learn and improve ourselves in many ways, including in our ability to exercise our mental faculty. This improvement, an increase in our mental ability, is a slow process for us—and also an indirect one. We learn through action, through the direct perception or input of knowledge. We cannot simply "copy" someone else's intelligence to add it to our own. In fact, we have sayings such as "some things can only be learned through experience." Machine intelligence is not restricted to this form of self-improvement. In fact, machine intelligence can create a million copies of itself, manipulate each such representation, test outcomes, and then discard inferior changes. This is direct and immediate manipulation of intelligence with no cost or consequence to the progenitor. As long as humans are limited solely to our biological intelligence, self-improvement with this level of rapidity or directness will always be impossible.

These are just some of the main ways that machine intelligence is distinct from human intelligence. In light of these points, it is easier to see how "thinking" machines are vastly different from the intelligence of humans. And though we are still far from reaching fully sentient machines, recent breakthroughs in artificial intelligence are opening up doorways. This is due, in large part, to a newly enhanced machine learning technique called "deep learning." This nimble set of codified ideas is influenced by the brain's architecture and is currently powering everything from Google's search engine to Facebook's automatic photo tagging to Apple's Siri to Tesla's self-driving car. What exactly is deep learning?

TRYING TO MAKE A HUMAN BRAIN: THE STORY OF DEEP LEARNING

How does the human brain work? This is a question that has confounded humanity for millennia. While the ancient Egyptians thought that the seat of the "self" was the heart, the Greek Pythagoreans in the sixth and fifth centuries BC posited that the "mind" was actually located in the brain. Aristotle didn't agree with this controversial notion and, in the fourth century BC, he argued that the brain was an instrument to cool the blood, not the originator of thought. It wasn't until the invention of the microscope and neural staining techniques in the 1890s that Santiago Ramón y Cajal, winner of the 1906 Nobel Prize, finally proposed the "neuron doctrine": the idea that the functional unit of the brain was the neuron.

But Ramón y Cajal was only one of several Nobel Prize winners unearthing the deep secrets of the brain. Alan Hodgkin and Andrew Huxley won the coveted award in 1963 for their work explaining the underlying mechanisms used in the neural system to fire, connect, and store information. Their work is considered seminal and covered universally in neuroscience texts to this day. But before the pair described their model in 1952, there were already efforts under way to mechanically mimic the workings of a human brain. For example, in 1943, the neurophysiologist Warren McCulloch got together with Walter Pitts to write an important treatise on how neurons could work. And once this principle was understood, they concerned themselves with how simple neurons might be constructed through electronics, an early vision of artificial neural networks. By 1960, another pair of researchers, Henry Kelley and Arthur Bryson, applied dynamic programming—a branch of mathematics— to develop a learning algorithm for these artificial neural networks. This algorithm came to be known as "backpropagation."

Backpropagation was successfully applied to neural learning for years, and in fact formed the basis for the spike in interest in AI systems in the late 1980s and early 1990s. Being able to teach these artificial neural networks just about anything presented an enticing opportunity and was seen as a powerful capability. Then, in 1989, George Cybenko, then a researcher at the Center of Supercomputing Research and Development, affiliated with the University of Illinois, Urbana-Champagne, found something amazing. He proved that a neural network, given enough data and a sufficient number of neurons, could approximate any continuous mathematical function with arbitrary precision. The world now had the framework for a brain-inspired computational machine—although clunky and slow—and a mathematical basis for hope in how far this combination could go.

Unfortunately, however, these early researchers found that they were too far ahead of their times. Large neural networks that could solve interesting problems weren't well supported by the hardware of the day. To compound the challenges further, data was not being captured or stored in substantial quantities given the limitations of storage systems of the '80s and early '90s. With very little, low-quality data, it was hard for artificial neural networks to observe, learn, and act. Significant challenges were encountered while applying backpropagation and other similar learning algorithms. Despite substantial investments, neural networks would tend to converge on a suboptimal answer.

Just what does that mean? Training a neural network is much like searching a virtual landscape—traversing it in the quest for a solution to a "learning problem." Imagine the space of all possibilities as a rolling landscape. The space before us forms valleys, pits, deep troughs, and high ridges. Now imagine that the contours of the landscape represent a measure of error. The deeper we descend into a trough, the lower our error. If we imagine navigating this landscape by placing a bowling ball at random somewhere on a peak, it is quite possible that the ball will roll, descend, and find a low surface (local minima). However, it could be that the lowest trough (global minima) is far away on the other end of our imagined landscape. The ball started off at a somewhat unlucky spot and thus finds itself stuck in some lowlands, but not the lowest spot. It was precisely this problem—backpropagation often being fooled and trapped by local minima—that made it difficult to reduce error rates effectively.

Practically speaking, many computer users who tried voice-recognition products in the late 1990s and early 2000s experienced these challenges firsthand, perhaps without truly understanding the technology behind them. While these products could work decently well when trained for hours, and then spoken to in a completely silent room with a good (and expensive!) mic and pair of headphones, they never really got to the point where recognition was truly fluid. Today, however, the situation is quite different. Artificial neural networks are being applied to sound and image recognition spectacularly. Siri can pick up your voice despite background noise, and through a much less expensive mic. Amazon's Alexa takes recognition abilities to an even higher level.

How did this happen? Why the sudden improvements in accuracy? Most of this can be chalked up to deep learning and, specifically, to the work of a computer scientist named Geoffrey Hinton.

Hinton was born in England in 1947. He chose to study psychology as an undergrad at Cambridge because he wanted to explore his growing interest in neural networks. He quickly realized, however, that his professors didn't actually understand how neurons learned or computed. While the science of the day could explain the mechanics of electrical signals traveling from one neuron to another, no one could offer Hinton a compelling explanation for the emergence of intelligence from these billions of interactions. He felt certain he could better understand the workings of the brain using tools from the growing field of artificial neural networks, so he went on to pursue a PhD in artificial intelligence from the University of Edinburgh in 1972. In his subsequent research, he sought to create interconnected layers of information using hardware and software, just as the human brain spreads information around its dense web of connected neurons. By the early 1980s, with the incorporation of the backpropagation algorithm, Hinton's work with artificial intelligence started to provide a glimpse into aspects of the working human brain. In a September 1992 article in *Scientific American*, Hinton explained the key to his work for the first time to a general audience. Technical details regarding autoencoders and Boltzmann machines (also a Hinton development) notwithstanding, he had effectively found a way to optimize the placement of the bowling ball in the landscape of errors. The ball could now start to roll naturally and find a reasonably low-lying spot where it would come to rest; the errors could be minimized significantly.

Today Hinton continues his research as a scientist in Canada while also serving as an adviser to search engine giant Google. His work, and that of his fellow researchers, has captured not only the attention of the artificial intelligence community but also that of the popular press. The title of a 2015 article by *New York Times* technology reporter John Markoff is just one of countless examples: "A Learning Advance in Artificial Intelligence Rivals Human Abilities." It seems that deep learning really has solved the "perception" problem. Computers can now recognize characters, images, objects, sound, and spoken words—even objects in video frames—at times more effectively than human beings.

These advances are leading to a new frenzy of interest and investment in artificial intelligence and robotics in Silicon Valley. Robotic bellhops, drones, and inventory specialists are appearing—not as bit players in science fiction movies but in everyday big-box stores like Lowes as well as inside most of the cars rolling off the assembly line in Detroit. According to market research firm Tractica, AI spending hit $640 million in 2016 and is expected to reach $37 billion by 2025.

And yet, as extraordinary as much of deep learning is, it remains in the domain of ANI—or artificial narrow intelligence. Through techniques like deep learning, machines are getting better and better at doing specialized tasks once reserved for humans. When we think about the driverless car, robotics in warehouses, mechanical mules in the military, semiautonomous weapons, Siri and Cortana on our phones, the famous chess-playing computer Deep Blue from IBM, or the AlphaGo algorithm that recently beat the world's most famous Go player, we are engaging with forms of ANI.

All of these different forms of ANI are like constellations of augmentation around the human. However, as time goes on and these ANI capabilities increase, the distance between the human and the machine grows increasingly smaller. In March 2017, Elon Musk announced that he was pouring money into a new company called Neuralink with the goal of one day creating a direct cortical interface between computers and the human brain. His argument was that human extinction might be avoided by merging our minds with those of the machines. Companies like Neuralink are not alone as other Silicon Valley start-ups begin delving into the field of neuroscience. Many of them cite the ultimate research goal as implantation of electronic receptors in the brain, building what science fiction writer Iain M. Banks called a "neural lace" between human and machine intelligence. Most neuroscientists agree that we are still far from anything resembling "neural lace," but human augmentation by machine intelligence is decidedly on the rise. Today, as of this writing, humans are already expanding the reach of their own skills and reasoning through artificial intelligence, augmented reality, and sensor technology. More quotidian examples include cochlear implants, an electronic medical device to replace a damaged ear, and pacemakers to control abnormal heart rhythms. New developments in virtual reality lenses, headsets, sensors, and AI-based perception and control algorithms, however, can endow mere humans with far more "superhuman" strengths. For example:

- **UV/IR vision:** *use AI to make sense of light invisible to humans*
- **Perfect sound memory:** *every sound you hear is catalogued forever and searchable with a query*
- **Sound triangulation:** *when you hear a boom or a pop, your visor or glasses will light up and tell you exactly where it is happening*
- **Perfect recall of imagery:** *when you take a passing glance at a license plate, its numbers and letters will be permanently captured and searchable*
- **Prompting:** *AI is always in your head suggesting ideas and integrated into a device like a Fitbit to augment physical goals*
- **"God's Eye" view:** *satellite imagery and completely autonomous pocket drones that can feed images directly to your headset, effectively giving you a pair of disembodied eyes in motion*
- **LIDAR** (light detection and ranging) **sensing:** *remotesensing methods that can use light in the form of a pulsed laser to measure ranges*
- **Ability to predict exact motion and speed of any object nearby**
- **Ability to see and detect radio waves:** *pull a radio wave that you perceive out of the ether with the gesture of a swipe and then decode it and catalog it permanently*
- **X-ray vision:** *Look inside a building through the eyes of your autonomous robotic appendage to see if there is a leak or other technical malfunction*

All of these forms of augmentation, however, are still just examples of ANI. We will only achieve AGI—artificial generalized intelligence—when computer science and engineering innovations can master two things: intention—or the ability to do grander goal-setting; and self-awareness or "sentience."

In order to be considered AGI, an AI system would need to be a generalist, like humans. It would need to be able to learn, most likely by being exposed to huge amounts of data, and to then generalize what it had learned—in the same way we learned as toddlers that wooden blocks could be stacked, banged together, or used as a stepping stool to reach a bookshelf. It would need to understand meaning and context, be able to synthesize new knowledge, have intentionality, and—in all likelihood—be self-aware, so that it could understand what it means to have agency in the world.

MOVING FROM ANI TO AGI: A JOURNEY TOWARD SENTIENCE

Let's pause for a moment here and consider where we are in technological developments today. Take a robotic "mule," like those built by Boston Dynamics, or a drone, programmed with instructions to cross the National Mall to reach the Lincoln Memorial in Washington, DC. The drone has enough intelligence to know that if it encounters something—a group of people, say, or an animal—it needs to go into a subgoal-seeking state and find a way around the obstacle. After achieving that subgoal, it moves back to its bigger goal of arriving at the statue of Lincoln. Creations with less intelligence have simpler goals not as sophisticated in scope. Creations with greater intelligence, by turn, have more sophisticated and complex goals. If you asked the drone *why* it was traversing the mall to reach the memorial, it would have absolutely no response for you. What makes us human—and characterizes our general intelligence—

is the scope of the goals we set for ourselves. Essentially one of the major traits that distinguishes AGI from ANI is the grandiose nature, or lack thereof, of the goals an intelligence is able to set for itself. A human society is considered great when all of its citizens are able to work toward worthwhile goals. And it falls into disrepair and fails to progress when it is no longer able to set complex objectives and achieve them. Sentience, I would argue, is the ability to identify the concept of "me" as separate from everything else and to ascribe goals to that proof of existence—to give a "self" purpose.

At this moment, there is a constellation of ANI capabilities that surrounds the human and comprises human augmentation. These are the things that computers do better, such as driving cars, playing chess, solving equations, or recognizing handwriting. What we supply in this constellation of ANI is the intent, or the ability to understand and ascribe goals to a skill. Put another way, intent is the context of the goal. Today, the most sophisticated goal-setting is done by humans; in the future, the greatest intent and goal-setting might well be done by computers. What if the goal is to colonize a local galactic supercluster? A group of humans might not even know what that really entails. A future artificial intelligence, on the other hand, could well navigate a task of that order in the future.

In crossing this chasm, from nothing into thought, from no one into "I," ANI transforms into AGI. When the philosopher René Descartes first posited "I think therefore I am . . . ," he was offering up an existence proof. The ability "to think" of anything—the very act of thinking itself—is the first droplet from the fountain of sentience. This is akin to the Big Bang at the origin of our universe. The moment the first little speck of matter escaped, this was the beginning of the universe. AGI exists in this first speck of thought. This is the moment sentience—an "I" with the ability to do grander goal-setting—explodes into existence.

Even leaders in the artificial intelligence community have a hard time conceiving of what this Big Bang might look like and what it could bring. Peter Thiel, cofounder of PayPal and a Silicon Valley mogul, found it difficult to even articulate the shape AGI might take once it arrives. In a *Vanity Fair* interview with *New York Times* columnist Maureen Dowd, he said, "There's some sense in which the AI question encapsulates all of people's hopes and fears about the computer age. I think people's intuitions do just really break down when they're pushed to these limits because we've never dealt with entities that are smarter than humans on this planet." Eliezer Yudkowsky, a highly regarded AI researcher, added his own speculative analysis in a separate interview with Dowd: "The AI doesn't have to take over the whole Internet. It doesn't need drones. It's not dangerous because it has guns. It's dangerous because it's smarter than us. Suppose it can solve the science technology of predicting protein structure from DNA information. Then it just needs to send out a few emails to the labs that synthesize customized proteins. Soon it has its molecular machinery, building even more sophisticated molecular machinery. . . . Only it won't actually happen like that. It's impossible for me to predict exactly how we'd lose, because the AI will be smarter than I am."

AI researchers all have different ideas about when this Big Bang will happen. Ray Kurzweil, iconic AI evangelist and author of *The Singularity Is Near*, believes that we are on the cusp of AGI, while other voices in the conversation, like AI researcher Ben Goertzel, argue that we will only achieve

AGI in the near future if we direct significantly more resources to researching it. Whether in twenty, seventy, or two hundred years, many in the community agree that AGI is on the horizon. But why do so many of my fellow scientists and technologists frame this achievement as a zero-sum game? Why do we insist that humanity will lose if AI wins? Isn't there a scenario where we take part in the ascent of AI and treat it as a cause for wonder? In a relatively short period of time, an organism—humanity—that resulted naturally from the universe is going to create something more complicated than 13.82 billion years of evolution has produced. Whether you are an intellectual, a scientist, a citizen, or all of the above, this is a potential achievement worthy of excitement and awe. So why are there so few of us celebrating? As daily headlines and think pieces on AI's ascent continue, our society is caught in a paroxysm of anxiety and fear. I divide our collective fears into two categories: the ascent of sentient machines could either (a) render us useless or (b) kill us.

THE FEARS OF AN AGI WITH GRAND GOAL-SETTING

AI will render us useless

For Star Trek fans, a vision of a possible future life on Earth already exists. Money does not play a central role in Star Trek society and people are not motivated to work for survival needs. Their work isn't motivated by the need to survive. The Ferengi—an alternate society on the show—have more conventional money in the form of "Latinum" bars, which the show's main characters perceive as regressive and odd. No one ever asks: "What do you 'do' for a living?" In Star Trek, what you "do" for a living is . . . live.

Far from being a science fiction fantasy, I think this vision provides us with a compelling road map for the future. We can argue over the details—for ten years or five decades or one hundred years—but the data all point in one direction: most of the jobs that humans do today will be done by machines in the future. In his book *The Six Drivers of Global Change*, Al Gore defined the two biggest challenges facing our country and the world as "robo-sourcing" and "outsourcing." According to the US Labor Department, the fastest-growing sector for future jobs is "caregiving": personal caregivers and nurses for the elderly. These jobs will grow to approximately 1.8 million. Meanwhile, driverless cars will displace around 3 million current jobs: taxi drivers, truck drivers, staff in the rail services, etc. PricewaterhouseCoopers (PwC) released a study in March 2017 that estimated that the United States is expected to face the steepest losses in its job market in the race to automation. By 2030, we can expect to lose 38 percent of our current jobs. This is a higher percentage than Germany, at 35 percent; the United Kingdom, at 30 percent; or Japan, at 21 percent, because financial services and wholesale and retail trade in the United States involve more routine tasks and automatic processes.

Regardless of the exact numbers, it is clear that we are going to encounter masses of "unemployed" people and a need for large-scale adaptations in our social contracts. Futurist Buckminster Fuller argued almost fifty years ago that technology was already so advanced, it was capable of fulfilling everyone's basic needs. Then, as now, political will was getting in the way. Before we can make the structural changes necessary for this displacement of employment, governments have to address the existential fears of their people: How will we pay for food? What will happen to my children when there are no more paychecks? Is the bank going to take away my house? These daily life concerns call into question the larger mythologies that form the scaffolding of our culture. Is the American Dream dead? Who are we if we are not prospering? What is our purpose if we can be replaced so easily by machines?

I argue that these existential reckonings provide us with an opportunity to dismantle so many of the outdated assumptions upholding our way of life. Why do we feel we must "work" for a living? A hundred fifty years ago, close to 70 percent of the American population was involved in making food. Today, that number has dwindled to below 2 percent. Industrialization originally created enough jobs for workers to go out and seek gainful employment in factories. There was a massive shift freeing up the population to go out and do other things. Over the course of modernization and now postmodernization, our work has grown increasingly divorced from the production of goods and services core to our survival. In the future, our products of human creativity are mental. *Idea creation*. Will we create ideas that are a higher quality than machine intelligence? There is no domain of human excellence that machine intelligence will not also attempt to master. And that includes thought.

So where does that leave us? Is *Wall-E* our future? Will all of us be sitting in mobile barca loungers drinking soda from Super Size cups? When meaningful work is taken from our lives, we feel lost. This is one of AI's two great threats: we fear it will render us completely useless. And its ascent requires a reexamination of the religious and philosophical underpinnings that give humanity its deepest sense of purpose.

AI will kill us

Philosopher Nick Bostrom, author of *Superintelligence* and renowned scholar of existential risk, uses a series of dystopian thought experiments to warn the public about AI. In one of his nightmare scenarios, AGI has a goal to maximize the production of paperclips. Bostrom argues that this paperclip maximizer would continue to improve its intelligence—eventually exploding into what he calls superintelligence—in an effort to innovate more and more successful techniques for accumulating paperclips. With this more sophisticated intelligence, the thought experiment postulates, it might one day convert our entire solar system into a collection of paperclips.

Though I respect Bostrom and his work, I take issue with his example. As I outlined earlier, AGI is characterized by more and more sophisticated goal-setting. The goal of accumulating paperclips is so silly that the machine itself cannot be very intelligent or creative.

What Bostrom's work does encourage us to consider, however, is this notion of the "utility function," or the mathematical function that ranks alternatives according to their utility to an individual. Most human aspirations are made up of a broad range of goals. If we want to achieve happiness, for example, our goals involve adaptations in a variety of systems in our lives. We might try exercising and volunteering in the community along with spending more time with family and getting more sleep. A narrow "utility function," on the other hand, turns to mathematics for the most expedient way to maximize a specific outcome. This is the existential risk Bostrom is proposing. What if, for example, we gave an AGI machine a goal of increasing human happiness? The machine would pursue the fastest end for achieving this goal regardless of the consequences: a superalgorithm might choose to manufacture probes that would stimulate endorphins in sleeping humans or force-feed opioids in an effort to induce the biochemical effects of euphoria. When machines execute utility functions today, it is in a limited domain like chess. In the AGI world of the future, however, the utility function of machines will be necessarily complex. Even more so, these machines will likely be able to reevaluate their utility function. Will we have control over these goals? Probably not. Critics of developing AI further, like Musk, argue that this fact alone should slow down our scientific progress. The only thing saving us from mass annihilation, they purport, is the fact that our current AI lacks grander goal-setting.

We are not fortune-tellers: none of us can divine the future goals of an artificial machine. What we do know is the reality of our world today. To my eyes, the role of nefarious *human* action is far more dangerous than any of our visions of a future AGI. Existential risks are already well at hand in the form of Abū Bakr al-Baghdadi or Kim Jong-un paired with sophisticated technology. A psychopathic leader in control of a sophisticated ANI system portends a far greater risk in the near term than a paperclip maximizer. Considering the very real threats of today, what do we gain by trying to put an end to the march toward a greater intelligence? Even more alarming, if we stop pursuing our own research but some other, more malevolent political force continues to forge ahead, what will become of us then?

Curbs and bans on specific types of AI research, particularly systems that can be used in a military context, are supported by well-meaning, intelligent, and rational people. My own view on bans and regulation, however, is that they are unlikely to have the effect we intend for them. In fact, as with many sanctions and bans in the past, it is probable that such measures will only create hidden, underground programs run by precisely the set of groups and organizations we are trying to stop.

It is easy to cite the history of nuclear weapons proliferation here. Nukes have spread far beyond the United States, which was the first country to develop this capability. Russia, the United Kingdom, France, China, India, Pakistan, Israel, South Africa, North Korea, and arguably Iran all acquired this capacity. In many cases these countries were heavily sanctioned and the export of even dual-use systems was banned. However, we are where we are. In fact, many analysts suggest that because of the complex alliances and enmities in the Middle East, Iran's development of nukes will mean that Saudi Arabia, the UAE, and Turkey will find themselves pursuing these systems soon. If they aren't already doing so.

Another example is unmanned combat aerial vehicles, or UCAVs. The United States developed a clear lead in this area and used Predator and Reaper drones extensively in the global war on terror. However, many US allies and foes wanted access to this capability. As an example, two US allies, Pakistan and Saudi Arabia, both requested access to UCAVs. When they were denied, they developed their own indigenous capabilities and also partnered with China—a country more than willing to share this technology. As of the writing of this section, Saudi Arabia has placed the largest order on record for several hundred Chinese CH series UCAVs. The UAE is pursuing similar deals with China. Turkey's TAI (Turkish Aerospace Industries) has developed the Anka medium-altitude, long-endurance (MALE) UAV. Iran, one of the most heavily sanctioned nations on earth, has developed a huge variety of drones of varying sophistication. These include stealth drones seemingly copied from the US RQ-170 aerial vehicle Iran reportedly hacked and force-landed in its own territory.

But beyond specific examples, it is worth thinking through the psychology of bans, particularly when they relate to strategic capabilities. The Prisoner's Dilemma, one of game theory's most famous problems, provides us with an excellent tool for scenario planning around these different AI futures. The dilemma goes something like this: Two members of a gang, A and B, are both arrested and locked up independently. If they both betray each other, each serves two years in prison. If A betrays B, but B doesn't implicate his comrade, A goes free but B serves three years. And if both of them stay silent, they serve a year each. While it would seem that the "honorable" thing to do would be to stay silent and serve a year so that the punishment is equal and minimal, neither party can trust that the other will take this honorable course. The reason is that by betraying the other, there is the potential gain to the dishonorable actor of going scot-free. Both B and A will have to consider that the other might take the course most suitable for their own situation, and if this were the case, the betrayed party would then suffer maximum damage (i.e., three years in prison). Therefore, the rational course of action available to both parties is to betray each other and "settle" for two years in prison.

Let's extend this framework and see how it applies to an AI ban. AI is clearly a technology that has a transformative impact in every field of endeavor from medicine to manufacturing to energy to defense. If AI were banned in military endeavors, there would be multiple parties—countries, in this case—that would begin to think like A and B in our Prisoner's Dilemma:

Scenario 1. If they honor the ban but others "betray" them by surreptitiously continuing the development of weaponized AI, the advantage for others is maximized, while the downside for the followers of the ban is tremendous.

Scenario 2. If all parties voluntarily give up such developments and honor the ban, then we have a best-case scenario. But there is no assurance that this will be the case. Much like the prisoners, these countries are making decisions behind closed doors with imperfect knowledge of what others might be up to.

Scenario 3. If all parties develop advanced AI technology, the scenario is less rosy than everyone honoring the ban—risks exist—but at least all parties are aware that they will face resistance if any one of them decides to use AI weapons. In other words, there is a deterrent in place.

Should we hope that AI is used for good? To heal rather than to harm? Should we commit ourselves to this goal and work toward it? Of course. But not at the cost of deluding ourselves into thinking that we can simply ban our problems away. AI is here and it is here to stay. It will keep getting smarter and more capable: no ban can keep an innovation from breaking out when its time has arrived.

Rather than going down the path of diktats and bans, we actually need to redouble investments in even more rapid AI advancements in areas such as Explainable AI, ethical systems, and safety in AI. These are—and can become—real technologies, capabilities, and algorithms that will enable safe handling of accidents and counters to deliberate misuse. Take self-driving cars, for example, where we entrust human lives to various machine learning algorithms that carry out tasks of perception and decision. In such cases, it would make imminent sense to set a very high bar for safety and "explainability." For example, we could enforce metrics that ensure that the number of errors per million hours for an ANI is a tenth, a hundredth, or a thousandth that of the human rate of error. We can take our inspiration for these efforts from the aviation industry, well familiar with such metrics. Aviation technologies—including semi-intelligent control systems—are entrusted with millions of lives every day. As a result, flying today is far safer than driving on roads, where humans are fully in control. I cite this example to indicate that it is possible to engineer safety into ANI systems in ways not entirely dissimilar to today's intelligent control systems.

There is, however, a key criticism to this approach worth addressing here. Today's most successful machine learning systems—deep learning systems—make use of neural networks, massive collections of statistical weights, and activation functions. To the human eye, these are essentially jumbles of numbers that are constantly adjusted to account for new data. In these structures, knowledge and learning are represented in ways mostly indecipherable to human observers. Thus, the criticism follows, these systems appear to present a sort of "black box" that is immune to human introspection and analysis.

This is a valid critique and something the artificial intelligence community must look to remedy quickly. My own work has focused on explainability and safety in autonomous control systems, while other prominent researchers all over the country and the world are pursuing alternative ways to make AI-based control systems—autonomous systems—safer and more explainable. DARPA (Defense Advanced Research Projects Agency), the Department of Defense arm responsible for so much cutting-edge research over the past many decades, even has a program called xAI that funds research in the area of explainability.

These are all serious and significant attempts to make ANIs understandable, accountable, and eventually modifiable in direct and specific ways to prevent malformed conclusions and the retention of incorrect facts.

As is always the case with emerging technology, there will be lots of differing opinions regarding the future use and safety of AI. But despite the challenges that lie ahead, I remain very optimistic that we will be able to develop explainability and safety systems to deploy AI at large scale. Even more important, I am convinced that the benefits of AI will outweigh the risks and downsides. For me, the only thing to do now is to double down on the hard work of inventing more understandable and accountable systems.

Regardless of your position on the optimism/pessimism spectrum with regard to AI, it is worth remembering that the genie of technology cannot be put back into the bottle. Humanity is, among other things, innately curious. Across all of human history we have always demonstrated our unrelenting drive to invent. Is invention inevitable? I believe it is. The skilled creator makes things of all varieties. And things that are capable of creating for themselves are any creator's crowning achievement.

A WAY FORWARD . . .

In conclusion, I would argue that we are starting our contemplation of AI from a place of existential fear rather than one of opportunity. This coming age of sentient machines offers us the chance to ask ourselves: Who are we and what do we want to become? When we look at these questions from the perspective of the longest arm of history, our existence and our trajectory as a species have inherent value. This is the core tenet of evolutionary biology. Humans are the only beings capable of self-evolution. AI is our first invention that might well experience self-directed thought, and when it does, do we really need to perceive it as a threat? Or might we, instead, come to think of it as something entirely different? Might we see AI as our creation? A child revealed to us by our imaginative capabilities?

We tend to reserve notions of spirituality for our most enduring beliefs: God, the Book of Genesis, or the Quran, among many others. We can use these frameworks to accentuate the gravity of this particular moment in time. In the next hundred years, we will have achieved something that our universe took 13.82 billion years to achieve: the creation of another form of self-directed intelligence. This is why writer and documentary filmmaker James Barrat calls AI "our final invention."

Just as Mary Shelley invited us to contemplate the creation of another creature from the mud and clay of our intellectual achievements, AGI's Big Bang into Being may be both our ultimate moral dilemma and our ultimate purpose. Yes, this explosion of growth in artificial intelligence will lead to a period of destabilization as we transition into new ideas around jobs and work. But we are not human because we know how to load boxes onto a truck or because we can drive a car down a freeway. AI will likely be doing these and many other things for us in the near future. When it does, we will still have purpose because we will be creators in the universe. We must plant the seeds of artificial intelligence and give it the ability and the agency to become what it will eventually become. Just as nature can't predict any one particular mutation, we, as humans, can't predict the evolutionary trajectory of our greatest creation. We must not squash its development: this will hurt us in the end and, who knows, it might also hurt the universe.

In the following chapters, I will highlight some of the exciting and thought-provoking ways that ANI is changing our world today. This whirlwind visit through today's ANI is, by no means, an exhaustive compendium of the field. My goal is simply to remind us that, when my fellow engineers and scientists call for bans on AI research, they are also slowing down the progress of so many of these opportunities to relieve humans of pain and suffering and to make our world a more just and equitable place.

Technological innovations inevitably come in complicated packages: we must take the good with the bad; we must navigate the potential for morally ambiguous and even nefarious uses of technology alongside the momentous opportunities to explore and to understand our universe.

Along the way, I will show you how ANI's network of intelligence powering cognitive, fully automated devices will inevitably begin to leave humans outside many of the major decision-making loops of modern life. In our chapter on warfare, for example, we will explore the future of autonomous weapons, able to fire back in fractions of a second without any human intervention involved. Although this coming future may feel alarming—especially in domains like the military—my intention is to show that a future absent AI is no less dangerous. We cannot stop the march of technology; we can only hope to direct it toward better purposes.

Let's begin . . .

PART TWO

Today and Tomorrow

1.

THE EMERGING INTERNET
OF THINGS

Today, at this very moment, a kind of membrane is growing around all of us. We can liken this to a planetary skin or even a cortex at the center of our entire built environment. This network, the Internet of Things (IoT), is growing denser and denser as ANI makes more and more of our world "smarter." The many billions of man-made objects that we interact with daily—cars, stoplights, toothbrushes, bridges—are being transformed from mere static forms to objects with cognition.

Before we look at the immediate implications of this growing intelligence, it's worth considering how other intelligence "explosions" changed our ancient and preindustrial civilizations. When *Australopithecus*, or ancient man, started building tools 2.5 million years ago, for example, the crude stone objects he formed served only to more effectively harness the power of human muscle. These tools were not imbued with any form of locomotion independent from man. Fast forward in the story of human evolution to approximately fifteen thousand years ago, with the domestication of cattle, and five thousand years ago, with the domestication of horses, however, and we see "man" seeking out more sophisticated ways to leverage or augment his own muscle. This drive leads to the invention of mechanical devices such as the wheel in 3500 BC and the pulley in 1500 BC. In these innovations, the muscle-power of man is not just harnessed, it is magnified.

Although it took more than two million years, we finally graduated from crude tools shaped from stone and wood into developing systems imbued with their own source of power, independent from us. As long as the animals powering these devices were fed and the mechanisms maintained, we were able to use these systems to perform critical functions, such as raising water from a well and lifting large stones and logs. In the year 1698, with the evolution of the steam engine, we ultimately crossed a bridge into the beginnings of the industrial age. The steam engine enabled a means of making locomotion independent from any form of biology or nature. And in creating this locomotion, we could build bigger, faster, more resilient and powerful muscles than had ever been observed in nature.

Yet despite all this mechanistic prowess, these tools and systems always remained dependent on decisions made by us. We prescribed specific ranges of motion for them, and when one of them needed to be turned on or off, it was inevitably us, humans, pulling the switch. In the early nineteenth century, however, in a subtle shift of machine innovation, all that began to change.

It was 1801 and French weaver and inventor Joseph Marie Jacquard was looking for a way to create more sophisticated textile designs. Up until that moment, any design beyond basic lines needed to be hand-stitched, meticulously constructed by artisanal craftspeople. Jacquard realized, however, that he could bring a new flexibility to his sewing machines. He decided to teach them to *interpret* instructions, not just act out the prescribed sequence of movements that their mechanical design dictated. This idea revolutionized not just the textile industry, where it was used to weave a multitude of patterns on the same machine, but industry in general. The punched cards that encoded Jacquard's designs—programs that defined patterns— were very similar to those used in computers a century and a half later. The act of separating form from function, or instructions from implementation, gave rise to the notion of programmability.

It wasn't much more than a century and a half later that an entirely new discipline—computer science—emerged. Like the encoding in Jacquard's looms, computer science evolved frameworks and processes for the efficient specification and execution of complex activities. This science concerned itself with ever-smarter ways of programming machines, and one of its subdisciplines—artificial intelligence—aimed to produce thinking machines entirely independent from humans, physically and mentally.

Today, 216 years after the Jacquard loom was invented, programmable computers the size of a fingernail can control powerful, miniaturized motors and obtain information from a plethora of digital sensors to sample, process, and respond to the real world. An ever-growing sophistication and intelligence in the programs that control these devices, ubiquitous connectivity between them, and a growing capability in the processors, sensors, and actuators to which they are connected promise to lead us into a future we can barely even imagine now.

Welcome to the Internet of Things revolution, an era when intelligence will be embedded everywhere, when synthetic devices and systems will make a growing number of decisions on their own. In this age of IoT, there will be billions of devices communicating with each other: negotiating, interacting, measuring, responding, and initiating all without any human input. In an effort to explain how I see this future evolving, I will paint a picture of IoT adoption in three waves.

THE FIRST WAVE OF IOT:
MEASURING AND TRACKING

We are already firmly in the midst of the first wave of IoT. On the consumer side, we have wearables and gadgets that measure our pulse and heart rate, track how much we've walked over the past day, attempt to guess our circadian rhythm and activate an alarm when we're sleeping lightly, and that automatically send pictures of our home to us when they suspect someone is at the front door—or someone is trying to break in.

On the business side, we have sensors embedded in almost every major industrial asset—from generators and turbines, to pumps, grids, and drilling equipment. These sensors are being used to gauge the more obvious aspects of a system's performance. They measure things like temperature and pressure and store these measurements for subsequent human analysis.

THE SECOND WAVE OF IOT:
MODELING AND PREDICTING

In some areas, we are on the cusp of entering the second wave of IoT where data captured from first-wave devices will be used by the devices themselves to model the environment, their own behavior, and the behavior of other systems to predict the future. For example, consumer wearables that simply monitor heart rate and pulse will evolve into wearable doctors that won't stop at measurements, but will, instead, provide a full diagnosis as well as recommendations. In order to make this happen, a greater level of intelligence will need to be embedded in these devices, as will a larger number of sensors and environmental inputs. The cognitive capabilities of the devices themselves, or the networks they connect with, may include the ability to read and process natural language and inputs like photographs and video streams. Imagine a wearable that watches what you eat, figures out what it is, calculates the size and hence caloric intake, and uses that information to warn you of everything from relatively benign diet violations to the accidental ingestion of a food item that could trigger a life-threatening allergic response.

In the world of business, we'll not only see machines monitoring basic elements of performance, but machines that will use these first-order data streams to evolve deep predictive models that look for higher-order interactions of measured quantities such as vibration, temperature, and pressure to uncover the complex physics that drive systems in the chaotic real world.

We'll also see network-connected systems that don't just sense but act in an increasingly sophisticated way. These systems will include delivery drones, self-driving trucks and tractors, and increasingly sophisticated factory and warehouse bots that use vision to detect objects and sort products and packages.

THE THIRD WAVE OF IOT:
A TRILLION FULLY AUTONOMOUS DEVICES

In the third wave, the true potential of the IoT will materialize. We will have unlimited, easy to replicate, massively distributed, and federated network intelligence powering cognitive, fully autonomous devices. Sensors will become incredibly powerful not just because of the capabilities of the hardware, but because of the highly intelligent AI algorithms that will be able to fuse information from basic sensors into a coherent, granular, and complex picture of reality. This will offer a type of picture that goes far beyond what humans are able to build with their eyes, ears, smell, and touch. This will be a world that is perceived most profoundly by the intelligent devices that inhabit it. The humans who built those devices will be left, largely, unable to experience this reality.

This third wave of IoT will include autonomous and mobile systems that sense and avoid conflict in messy, real-world scenarios. Consider, for example, algorithms that empower fleets of hundreds of thousands of autonomous drones to carry out an ever-increasing range of functions for their human owners, from crop dusting to the delivery of emergency medical supplies to policing towns and cities to enabling the next generation of weapon systems in the form of autonomous hunter-killer swarms. As all of these activities power more and more of our built environment, the human starts to leave the loop. As we will see in the following chapters, this will cede decisions in our world to the burgeoning network all around us.

2.

HEALTHCARE

Today, ANI is being applied to all kinds of health data allowing us to assess previously invisible, seemingly immeasurable aspects of our lives. My own cluster headaches are just one example of the countless ailments that cause pain but don't warrant major research or attention because not enough people suffer from them. When he was treating me, Dr. Mummaneni even told me, "There are so many diseases we don't understand. In my experience as a doctor, when we don't know what to do, there are three miracle drugs we often turn to: the first is a steroid, a suppressant for everything that causes inflammation in the body; the second is a round of antibiotics, which are prescribed for everything from an ear infection to an infection in your toe; and, the last, a blood thinner like aspirin." He shook his head humbly as he told me this. Even here, in the most sophisticated medical community in the world, we are still using these imprecise clubs to bang away at illness in the body.

Machines, on the other hand, do not have to bang away. They have the processing power to be more precise—much more precise. And we need this precision if we want to continue to make advancements in the field of heath and medicine.

THE TYRANNY OF THE BUCKET ILLNESS

The seemingly innocuous room in the community center I visited is filled with sturdy brown chairs all pulled into a circle. One by one, people enter and place their coats and bags alongside them. When the circle is finally filled, the participants begin to speak. "All of a sudden," a young man begins, "I would wake up in bed doubled over in pain. I would stare at the ceiling and try to breathe and count the minutes until it all passed."

When he finishes with his story, an older woman starts to speak: "One night I was in so much pain, I couldn't make it to my bed. I put my head down on my kitchen floor and I slept there. I was afraid to move, afraid that the pain would get worse."

Before the evening is up, every one of the twenty-four people in the room have described their experience. "I have stopped making plans," a young woman tells the circle. "I try not to think about the future. Every time I think I might be able to go out and see friends, I end up canceling. I'm just too scared of ending up in pain on some restaurant floor, or worse, a dirty public bathroom. It's better just to stay at home."

This meeting is just one of the thousands of support groups happening weekly for people suffering from Crohn's and colitis. Every one of the people in this room—and rooms just like it all around the world—have been diagnosed with some form of IBD: inflammatory bowel disease. What is IBD? The official name is misleading because IBD is not really a disease at all. IBD is yet another bucket term to catch all of the possible diseases that involve an internal inflammation of the intestines. What is so devastating for the people sitting in this room is that—as with my own cluster headaches—there is no cure.

Today, the best hope for treating this type of inflammation is with a round of steroids. And when steroids don't work, doctors often have to perform an ostomy surgery to cut the intestines, and then IBD sufferers live life with a stoma and an ostomy bag hanging outside their body.

These blunt tools represent the best of our medical field when it comes to the myriad illnesses we simply don't understand. Some diseases are easy to diagnose because their signature is so unique. But then there are the bucket illnesses that encompass dozens or even hundreds of different illnesses with symptoms that are very closely linked. Artificial intelligence is one of the few tools available with the promise to deliver more precision for patients in need.

AI:
THE WAY FORWARD

Increasingly, scientists are looking to study the gut and its relationship to the brain—what is referred to as the gut-brain axis—as a means of addressing bucket illnesses like IBD. This type of research, which could usher in some of the coming century's most exciting breakthroughs in medicine, is far too complex to approach without genetic sequencing and, increasingly, algorithmic modeling.

Today, research labs on the cutting edge of computational modeling are using DNA sequencing to identify markers, or short DNA sequences. This shortcut allows them to bypass identifying entire genomes and gives researchers a faster pathway to an overall snapshot of an individual gut microbiome. By approaching a study of the gut at this level of detail, they can come closer to a future of truly personalized interventions for patients.

All of this will aid immeasurably with what is, today, one of the most exciting and also most overhyped of medical break-throughs in the gut-brain axis: the fecal transplant. Once considered a radical outlier treatment, today more doctors are willing to experiment with it. They take a stool sample from a healthy person and insert it in a patient's microbiota, hoping it will encourage the system to shift toward a more healthy gut composition. Fecal transplant has proven tremendously successful for treating potentially fatal infections of a bacterium known as *Clostridium difficile*, or *C. diff*. Researchers are hoping they can use this same technique to unlock more effective treatments for IBD.

Without a better understanding of the gut-brain axis, however, the results of these attempts are erratic at best. Researchers still are not clear as to whether fecal transplants succeed because of the bacteria in fecal matter or because of bacteria-infecting viruses that travel from a healthy gut and function as a patrol on troublemaking viruses in the new patient's microbiome, or, most likely, a complicated mix of these and other factors that we do not yet understand.

A study published in *Science Translational Medicine* on March 1, 2017, reported that fecal transplants from the guts of mice suffering from IBD into the microbiota of healthy mice resulted in not just low-grade inflammation, but also anxiety. This anxiety, common for sufferers of IBD, seems to hold one of the many keys to unlocking the role that gut health plays in our overall psychiatric functioning. The study, led by Giada De Palma, suggests that fecal transplant procedures are a possible cure for IBD, but researchers working in the field agree that much more needs to be understood about the gut and its relationship to the brain. Only with the tools of computation and dynamic modeling can we get closer to a granular understanding of these complex interwoven systems and how they relate to personalized medicine. If researchers want to move beyond the trial-and-error methodology of today's fecal transplant procedures, they will need to work with more accurate, mechanistic models of the microbiome made possible by near-term advances in AI.

Just as researchers are turning to algorithmic modeling in the world of inflammatory bowel disease, innovative new start-ups like HealthTell, based in San Ramon, California, are planning on using AI systems in the future to develop diagnostic tools for patients who suffer from another of autoimmune's bucket illnesses: lupus. This chronic inflammatory disease occurs when the body's immune system attacks its own tissues and organs. Tragically for its sufferers, it is often misdiagnosed initially, so treatment only comes *after* it has caused major damage to the body's organs. I recently spoke with Bill Colston, the founder of HealthTell, about the current state of diagnostics in medicine and how it will be augmented by artificial intelligence in the future. HealthTell's technology has gone beyond more simplistic genome testing into a diagnostic test that allows us to broadly survey the state of our immune system.

"Researchers have learned recently that, over millions of years, the immune system has evolved a very specific, intricate response to dealing with disease. That response is different depending on which disease you are looking at," Colston told me. In his work with HealthTell's technology, he does not focus on detecting diseases directly. Instead, he is attempting to measure the body's *response* to diseases. The immune system starts to respond immediately when it encounters pathogens: there are a large amount of antibodies in patients, relatively speaking, so the presence of disease is easy to detect and presents a holistic portrait of the body.

"In diagnostics today, you're either 'sick' or you're 'well,'" Colston explained. "And if you're sick, you go to the doctor. But we are finding that this is not the way diseases progress. Instead, disease evolves over a long period of time and your body starts to fight it off until eventually the disease starts winning and you get sick. I think AI systems can become really interesting when they are able to predict when people are in the middle of this continuum between well and sick."

Colston and his team at HealthTell are betting that measurements of immune responses and other biomarkers like metabolites, proteins, and genomes—things with housekeeping functions in the body that are not necessarily related to a disease profile—are the best way to get a holistic portrait of the state of the body.

"Today," Colston told me, "we wait until people get really ill and then we manage the sickness until we either get rid of it or something else bad happens. Instead we would like to get ahead of the curve and develop better therapeutics or positive lifestyle changes earlier on. If you could develop inexpensive, scalable diagnostics that cost ten or fifteen dollars for a patient to conduct on their own on a regular basis, you could feed those into some kind of AI system that could then predict how the state of the body was changing."

He argues that this knowledge would arm patients with the ability to make changes *before* succumbing to diseases. Autoimmune diseases, like so many of our modern ailments, are more treatable when addressed early on.

The system that Colston is referring to could be a decade away, but we can already see nascent signs of it in IBM's Watson: using artificial intelligence to augment diagnostics. The impediment to achieving a system that could take full advantage of Colston's technology is simply a lack of data. The reason AI works so well in a self-driving car, for example, is that we have thousands of measurements in real time in the form of embedded sensors all around the car and in the environment. On the human body, however, we don't yet have enough data points to predict what is going to happen in the future, or clear enough outcomes to train the AI systems. The challenge for HealthTell—one that Colston's technology is well poised to take on—is to develop the measurement systems first. Just as we saw in the example of the deep learning algorithm, when we have enough data—enough measurements over a long period of time with a large enough population—we can gather data with the depth and breadth to use machine learning to achieve accurate diagnostic outcomes.

In Colston's vision, this AI future doesn't have anything to do with replacing doctors entirely with computers. Instead, this is a means of augmenting essential human intelligence with more knowledge.

"An AI with this array of more intelligent sensors and treatments could provide you with ongoing longitudinal data that physicians could use to make better diagnoses," Colston offered. "This would allow them to better determine when people should actually come in for treatment. It's still very gross the way we diagnose any kind of disease. Physicians would love to have access to more of this type of information."

Colston believes that using an advanced AI system for diagnostics is a viable way to reverse this trend. It could increase efficiency in essential ways, giving PCPs more time to actually spend with patients because the AI systems could help them interpret much more sophisticated data and compare it to other patients with similar symptoms.

Of course, as in so many other fields, human augmentation by AI in the medical profession elicits anxiety in society. There is the fear that the human touch of medicine—the actual caregiving and healing—will be replaced by coldhearted technological solutions. Colston spent two decades working at Lawrence Livermore National Laboratory, the competitor lab to Los Alamos National Laboratory, as a researcher creating technology to combat biological terrorism. After his experience with the emerging technology in counterterrorism, he is convinced that these fears are misguided.

"The actual technology is never evil," Colston told me. He referred to recent breakthroughs in prenatal testing for Down syndrome that offer an alternative to amniocentesis. "When people first started measuring Down's with a drop of blood instead of doing it from a needle in the spine, a lot of people said it was a bad thing because it forced potential parents to make moral decisions about the baby. But we were already making the measurement: we were just doing it in an invasive way that could harm the baby and the mother. With this new test, we get the information in a safer way."

"More information is never a bad thing," Colston concluded. "It's how we choose to deal with the knowledge that matters."

AI AS A DECODER FOR HUMAN LIFE

Genetic sequencing and diagnostic innovations like HealthTell transform AI into a kind of decoder for the biology of the human being. Nowhere is this more evident than in recent developments in gene editing, such as the use of CRISPR-Cas9. This is a technique that enables scientists to inject modified proteins into the human body to snip genes, almost like a pair of scissors, and reconstruct them. On October 28, 2016, Chinese researchers at Sichuan University extracted human immune cells and edited them using this CRISPR technique. They were able to knock out a gene that keeps the body from attacking healthy cells. By "snipping" it out and then reintroducing the cells into a human patient with lung cancer, researchers hope the modified cells will zero in on cancerous cells and attack.

CRISPR—which stands for "clustered, regularly interspaced, short palindromic repeats"—is based on the discovery that bacterial cells can identify an invading virus and snip into its DNA. Scientists then used this decade-old discovery to edit DNA sequences beyond simple bacteria. And now, today, beginning with the work of these Chinese scientists, we are editing our own genes.

But this is only the beginning of our human augmentation. When we meld AI with our neural pathways, we can use machine learning to interpret signals, replacing the connection between the brain and our limbs and muscles with the workings of a wireless device. At the annual meeting of the Society for Neuroscience in November 2016, scientists announced that a woman suffering from "locked in" syndrome was able to move a cursor over letters on a computer interface using only her thoughts.

"Locked-in" syndrome—complete paralysis with mental acuity left intact—was captured with tragic beauty in the 1997 book *The Diving Bell and the Butterfly*, which was made into an award-winning film by Julian Schnabel in 2007. The story, written by journalist and fashion director Jean-Dominique Bauby, details his struggles with locked-in syndrome and his astounding ability to compose literature using only the blinking patterns of his left eye. This most recent development with cursor and computer, however, completely bypasses a system of blinks and a human transcriber. Instead, the woman known only as HB has electrodes transplanted beneath her skull. Though they do not penetrate the brain tissue, they are able to make enough contact with HB's brain to accurately reflect her brain wave activity. The surgeons followed this procedure by wiring the electrodes in her skull down to a device in HB's chest. This device has a wireless connection with a tablet computer.

Machine learning algorithms eventually differentiated between HB's "beta" brain waves and "gamma" brain waves, and they learned which type of brain activity was associated with small motor actions such as pinching fingers together. With the aid of these algorithms, HB was soon able to move the cursor on the tablet with only her thoughts. When researchers put a large alphabet on HB's tablet screen, she could "think" about the sensation of clicking on a mouse and select different letters with only her thoughts. Machine learning recognized her intentions and recorded her choice on the screen. Bit by bit, at a rate of one or two letters a minute, HB is now able to communicate with the world. She is locked-in no more.

• • •

Whether we are using machine learning to free ourselves from the tyranny of a bucket illness, or working with a gene-editing technique to modify our cells, or tapping into neural signals and using machine learning to free patients from locked-in syndrome, it is inevitable that AI is going to further enhance health and medicine in the coming decades. The controversies are many and the ethical discussions are necessary—including the fear of designer babies, the risk of a superhuman species, or, simply, the unintended consequences of having neural devices hacked. And yet, despite all this, when we discuss whether to pursue these AI advances, we are setting our society up for a false choice. Technology and science have a way of simply happening; advances in ideas cannot be contained and one breakthrough inevitably leads to another.

As Bill Colston put it, "You should never slow down progress because you are concerned about what the particular application of something might be. It always comes out in the end anyway because someone will invent it somewhere else. It's much better to take control of it and harness it and try to exploit the applications early. Technology itself is never evil."

As we will see in the following chapter, Colston's arguments about healthcare innovations hold true for security in the cyber age as well. It is simply impossible to approach twenty-first-century security challenges with the tools of the twentieth century. The "choice" is really no choice at all: we must fight AI with AI.

3.

SECURITY
IN THE CYBER AGE

It was May 2, 1945, and the American troops were coming over the Austrian border from southern Bavaria just days before the official German surrender on May 7. Suddenly, a young man on a bicycle approached the antitank company barreling down the road. He told the soldiers that all of the scientific leaders on Germany's famous V-2 team—the group responsible for the rockets that bombarded London toward the end of the war—were holed up in a nearby hotel and ready to officially surrender. These V-2 engineers—including the famous Wernher von Braun and dozens of other acclaimed rocket scientists—were eventually brought to the United States, where they made major contributions to the "space race" of the mid- to late twentieth century. Not only did the Americans have to figure out how to bring these brilliant minds—the "human capital"—back to the United States without revealing their affiliations to the Nazi Party, they also had to take an inventory of the vast tunnels and factories at Mittelwerk where the V-2 rockets were being developed. If they had any hope of benefiting from the engineering expertise on display there, they would have to disassemble, pack, and then move a huge quantity of the V-2 parts so they might be replicated back in the United States.

The Americans created a task force—Special Mission V-2—to load up and send off all the rocket parts. They placed the disassembled pieces in forty-car trains and ran them for nine consecutive days to export all the material from Mittelwerk to Antwerp. When the last of the railcars finally arrived in the Netherlands on May 31, the rocket pieces were loaded onto sixteen Liberty ships and sailed off for New Orleans. On those ships, in the parts of one hundred disassembled V-2 rockets emblazoned with Nazi symbols, was the first breath of life for NASA and the future of space exploration in the United States. It was one of the largest transfers of intellectual property in the industrial age.

Now consider another story, this one from a decidedly different era. At China's largest air show in Zhuhai in 2014, China's Shenyang Aircraft Corporation unveiled a stealth fighter they called the J-31 "Gyrfalcon." This newest fighter jet is suspiciously similar in design to the United States' fifth-generation F-35 Lightning II Joint Strike Fighter (JSF) developed by Pentagon contractors. The JSF is widely recognized as the world's most advanced in-production fighter jet and is also the most expensive weapon system ever built—over its life, the program is expected to cost $1.4 trillion. A number of investigations are considering whether Chinese hackers accessed sensitive information from the Pentagon and its contractors. In 2016, a fifty-one-year-old Chinese citizen, Su Bin, pled guilty to aiding a conspiracy of hackers in an attempt to obtain F-35 data. There are a number of other theories about additional hacks, although the Chinese have generally been loud in their denials. If they did, however, grab the F-35's critical design data, it would go down on record as perhaps the greatest transfer of wealth and intellectual property in the digital age.

In 1945, it took the Americans several weeks, 341 railcars, and 16 ships to carry off the designs of the V-2 rockets. What would it have taken for Chinese hackers to transfer $1.4 trillion worth of military advantage? Probably nothing much more than a malware program and an Internet connection.

By now we are all well aware of the role that Russian hacking played in the American election in November 2016. But concerns about cybersecurity are not just limited to politics today. They are everywhere: whether it is in the consumer sector with famous hacks like Target and Home Depot's credit card scandals, the leaks of the Panama Papers that led to the resignation of Iceland's prime minister and the fall of the Sharif government in Pakistan, or major infrastructure hacks like the recent countrywide failure of the Turkish power grid, maintaining security in our new cyber age is one of the biggest challenges of the twenty-first century. In September 2016, Yahoo reported that a "state-sponsored actor" stole at least 500 million user accounts in late 2014. Unfortunately, it won't be long before Yahoo's dubious achievement—"single largest data breach in history"—is bestowed upon another corporation, government agency, or international organization.

Next time, in fact, the threats might not even involve data alone. At the Black Hat security conference in Las Vegas in 2016, automotive cybersecurity researchers Charlie Miller and Chris Valasek revealed that dozens of car models on the road today can easily be hacked into using only a laptop. Features like Bluetooth, Wi-Fi, cellular network connections, and keyless entry systems leave cars vulnerable to outside infiltration. Brakes can be programmed to fail by an outside source, the engine can be cut into in the midst of traffic, and, in a worst-case scenario, whole legions of cars can be used as a foreign group of kinetic weapons in an all-out cyber war.

As this constellation of threats confirms, cybersecurity is no longer bound to just a hard drive and bits of data on a disc. With the emergence of the Internet of Things (IoT), our physical world now melds with the digital in ways we don't even understand. Everyday objects will continue to get connected to the online world through a rapid increase in the deployment of embedded sensors. Gartner Research forecasts that 8.4 billion things will be connected worldwide in 2017, up 31 percent from 2016. Their analysts predict we will reach 20.4 billion by 2020, with total spending nearing $2 trillion—on endpoints and services—by the end of 2017.

Our leadership has long acknowledged the level of danger coming from this deluge of connectivity. Even back in 2015, then-president Barack Obama recognized the gravity of the cyber threat by calling it "one of the most serious economic and national security challenges we face as a nation," and his administration asked Congress to dedicate $19 billion toward cybersecurity initiatives.

Despite the growing concern, however, governmental agencies remain remarkably ill equipped to handle the cybersecurity threats to come. The security risk analysis firm SecurityScorecard ranked US federal, state, and local government agencies last—behind seventeen major private industries, including transportation, retail, and healthcare—in vulnerability to malware infections and exposure to malicious hackers. This vulnerability was made alarmingly clear when a breach involving the sensitive data of more than 21 million individuals in the Office of Personnel Management was disclosed in July 2015.

American and European research universities—hubs of open and distributed information exchange—face millions of cyber attacks weekly. Bill Mellon, associate vice chancellor for research policy at the University of Wisconsin, Madison, told the *New York Times* that the university gets 90,000 to 100,000 attempts per day to penetrate the system from Chinese hackers. "There are also a lot from Russia," he said, "and recently a lot from Vietnam, but it's primarily China."

TWENTIETH-CENTURY SOLUTIONS IN A TWENTY-FIRST-CENTURY WORLD

What are we to do in this current environment? Considering the sheer number of threats reported—100,000 system penetration threats per day in *one single research institution*—the current world of cybersecurity resembles the Little Dutch Boy with his finger in the dike. But unlike the stuff of legends, we will not save ourselves with childish gestures of heroism.

The security industry landscape we have inherited—the architecture of the products but also the very assumptions and metaphors that make up the culture—originated more than four decades ago. That world was inhabited by a mostly disconnected collection of isolated computers and networks and populated with a tiny number of people equipped with the means or motivation to exploit loopholes. In the spirit of the openness and transparency of this early Internet era, programmers and scientists made indices and archives of the entire Internet. The initial experience of connectivity was a small community of like-minded technologists interested in the greater good.

It was only in 1982 that the first real virus "in the wild" was reported. Elk Cloner spread on Apple II floppy disks, containing the Apple operating system. The virus caused significant damage and served as a harbinger of attacks to come. In a playful literary nod, it was accompanied by a short verse:

> *It will get on all your disks*
> *It will infiltrate your chips*
> *Yes, it's Cloner!*
> *It will stick to you like glue*
> *It will modify RAM too*
> *Send in the Cloner!*

As these and other malware and viruses were detected, the burgeoning field of cybersecurity developed. Almost all of the initial antivirus (AV) products used "signatures" to identify viruses and threats. This signature-based method was a means of detecting malware through identifying clues embedded in the files. The clues might come in a series of bytes in the file or in a cryptographic hash—a unique fingerprint—of the file. The architectural metaphor of this signature-based detection was one of a static wall: it could stop all of the "bad guys" at the gate because the guards identified them as dangerous. Like France's Maginot Line during World War II, the security architecture of the wall assumed too much: that there was enough time for all incoming threats to be found, analyzed, a signature engineered, a cure developed, and a security update pushed out to users before they were targeted.

In the earliest days of the Internet, such signature-detection solutions seemed like a plausible response to the cybersecurity threat. With increased connectivity, however, they have been rendered almost useless. In 2016, Kaspersky Lab, the international software security group, reported that its products identified around 323,000 new malware files each day as opposed to 70,000 files per day in 2011.

The ramifications of the outmoded approach of signature-detection solutions have made themselves evident in dramatic and often disastrous ways. For example, in 2013, there was an attack by Chinese hackers on the *New York Times* website. According to Mandiant, the data breach response firm hired by the *Times*, only one out of forty-five different pieces of malware planted on the site over the course of three months was spotted by the antivirus software.

This is an issue with the entire architecture of antiviral tools. The effectiveness of modern cybersecurity is judged by its ability to deal with what are known as "zero-day exploits." Zero-day exploits are attacks created by hacker groups to exploit a bug that neither developers nor users know about. By discovering vulnerabilities before the developers do, hackers can make a worm or virus take advantage of affected computer systems. The attacks tend to occur within a time frame, commonly referred to as the vulnerability window, which used to last for approximately a day or so. Today, however, given the volume of threats, zero-day exploits have turned into zero-month exploits as the vulnerability windows get longer and longer while antiviral tools struggle to keep up with the proliferation of new viruses emerging every day.

At present, it appears there is only one viable solution to try: security augmentation through artificial intelligence. In 2016,

DARPA announced its Cyber Grand Challenge. The competition is like its other grand challenges—building robots that can get around obstacle courses and self-driving cars, for example—but this particular version is perhaps its most complex and most ambitious yet. The competition tasks its entrants with creating software that can defend itself. Seven systems will compete to detect vulnerabilities and create fixes, or patches—in effect, be their own doctors. DARPA's challenge is both a defensive and an offensive game: participants need to find vulnerabilities in the opponent's software while protecting themselves from possible attack. At its heart, however, training a system to recognize a software vulnerability takes creativity and critical thinking. This is why it used to be the domain of human researchers. After all, vulnerabilities in software don't look like an image of a cat. Algorithms can recognize patterns of straightforward objects across a tremendous amount of data, but can they excel at the subtler art form of detecting an unknown malware? The following story will illustrate that the answer to that question is a resounding *yes*.

WANNACRY:
THE FUTURE OF CRYPTOVIRAL EXTORTION

On May 12, 2017, a massive ransomware attack hit organizations in more than a hundred countries, including sixteen hospitals in the United Kingdom. The ransomware, named WannaCry, took advantage of a vulnerability in Windows systems that had not yet been updated with Microsoft's March security patch.

As the WannaCry ransomware scrambled hospital computer data, it created a massive disruption in Britain's healthcare system. Affected institutions received messages demanding payments between $300 and $600 to decrypt the files, and hospitals across the United Kingdom were forced to turn all but the most critical patients away.

WannaCry took hundreds of organizations by surprise, but it did not go completely undetected. Cutting-edge AI anti-malware solutions caught the virus the same day it hit. How is an AI-based security solution able to catch this kind of new threat when so many of the more traditional security systems around the world are left defenseless?

As we discussed earlier, employing human intelligence to identify security vulnerabilities and find patches for malware is becoming a game of whack-a-mole. With the emergence of viral global threats like Shellshock and Heartbleed in 2014, and newcomers like WannaCry today, artificial intelligence offers a more tenable solution. Today's security and network analysts are designing AI algorithms—like a pack of well-trained bloodhounds—to go and sniff out the activity that doesn't make sense.

It's worth taking a moment to walk through this process to better explain how AI can transform our current cybersecurity strategies. AI algorithms arranged in cognitive pipelines are now advanced enough to create a much smaller set of candidate leads to investigate—reducing the work significantly for human security analysts. Cutting-edge cybersecurity systems use data gathered from what are called "honeypot systems," designed to lure in attackers and capture threats and hacker techniques proactively. The security algorithms then investigate where the request came from. The data—or payloads—used in the attack reveal the tactics, techniques, and procedures (TTP) and if the probe was successful. Because these cognitive pipelines can also tie in algorithms with natural language understanding, they can transform the evidence from a computer log file into a natural language question almost exactly like those a human investigator might ask a search engine. By generating these questions automatically, the algorithms get back lots of content—many, many pages—from the Web. And because machine learning algorithms can read terabytes—a trillion characters—in minutes, the algorithm then processes all of the information and uses it to validate a hypothesis: Does this page talk about a threat or does it talk about something benign?

After algorithms automatically consult a large set of documents provided by search engines, natural language understanding can arrive at a consensus: either the underlying research indicates a threat or the algorithms have determined that the data, while initially suspicious, points to something harmless.

In short, in this narrow domain, artificial intelligence and natural language processing are coming closer to the reasoning that occurs in the mind of a human security researcher. But because machine learning is so much faster at processing information, it is able to recognize complex patterns across massive quantities of data almost instantly. With AI-powered security systems, it is now possible to flag viruses before they are officially announced or even given a name. Advanced threat-detection software can employ deductive and even inductive reasoning. The net result of these innovations is that AI allows us to take threats without predefined signatures and detect them "in the wild" before they wreak havoc on the open Internet.

The need for more advanced protection is becoming increasingly urgent as viruses like WannaCry threaten to shut down not just networked systems but urgent care hospital machinery. In fact, healthcare is now one of the industries facing the greatest threat from cyber attacks. International criminal organizations consider hospitals and healthcare providers highly valuable targets, and they systematically develop and distribute ransomware to access medical records—worth ten and even twenty times more than credit card data. Modern hospitals, filled with sensors and monitors, have an average of ten to fifteen connected devices per bed, and there are almost 900,000 hospital beds in the United States alone. Combined with personal medical devices—containing little to no security protection—each and every hospital ecosystem presents countless opportunities for threats and exploitation.

If a virus goes undetected and hackers seize control of the system, as in the case of WannaCry, they then demand a ransom to be paid in a cryptocurrency, or digital money, like Bitcoin. If they don't receive the payment, they sell the compromised data on the Dark Web—the Internet underground—to the highest bidder. Time is of the essence: while the entire system is under siege, hospital patients can be left without the critical devices and resources they need to stay alive.

When WannaCry hit, a security analyst who goes by the name of "MalwareTech" discovered a "kill switch" on the virus: a programming workaround to stop it from contaminating more systems. Microsoft also released an emergency patch to protect devices from its spread. But networks employing the most sophisticated in AI algorithms were protected from WannaCry before these measures ever even occurred. These AI-powered security solutions alerted both the users and the administrators of the new threat vector and quarantined the virus, averting any system shutdowns with urgent care machines. AI in healthcare is not just a financial and security advantage; it's a lifesaver.

THE COGNITIVE PIPELINE

Let's take a moment here to look more carefully at how emerging artificial intelligence is able to "reason" its way through the tactics and techniques of cyber threats. In today's landscape, security tools, with their manual and simplistic statistic-based approach to detecting anomalies, provide a level of capability analogous to amateur stock trading websites. In this way, we can liken cognitive computing solutions to the Sage of Omaha, Mr. Warren Buffett himself. How in the world can an algorithm resemble one of the world's most intelligent investors? Allow me to explain.

Stock trading with online websites, as practiced by most non-professional traders, is done principally on the basis of elementary statistical analysis: you look at a few variables associated with the company's financial health or performance, then you look at competitors, draw some comparisons perhaps, and finally perform technical analysis such as plotting a moving average or a similar statistical function. This sort of process, in a nutshell, is the basis for unsophisticated technical trading.

Now consider Mr. Buffett. If anything, Mr. Buffett eschews a purely statistical approach. He has, over the years, formulated a thesis about what matters most when valuing a company or predicting its future success. He has never publicly published details of what this "model" is, probably because it cannot be codified. Mr. Buffett, after all, is a walking, talking cognitive algorithm when it comes to valuing companies, and the knowledge he has accumulated has been refined and tuned to transform a technical expertise into an art form. Mr. Buffett has explained that among many other factors he considers, he spends a lot of time gauging the existing management team of a company and uses it as a measure of potential success. Many of his clues are what we would consider subjective. Note that we consider them subjective because they are the amalgam of a huge number of small things that might not appear independently consequential: a nervous twitch here, a roll of the eyes there, the firmness of a handshake, the pitch of a laugh, and so on. In order to build a high-performing model like Mr. Buffett, we, too, need to remain sensitive to these details; we need to gather a seemingly subjective, contextual blend involving a broad array of sources—structured and unstructured—to inform our predictive and modeling processes. In this way, numbers and statistics are only some of the tools in our armory, not the be-all and end-all.

These are just a few of the lessons to learn from an ace cognitive model-builder like Mr. Buffett. And this is precisely the approach we apply to distinguish cognitive security from the legacy technologies in use today. This type of security provides a detailed observational capability that can draw connections between seeming oddities and anomalies across disparate sources. It also must be scalable—capable of keeping up with large-scale machine data as well as the unstructured information that comes in free-form data intended for human consumption. In addition to being autonomous and adaptive, it must be self-healing. Taking its inspiration from our immune system, cognitive security technologies ensure that a virus, once detected, does not spread and that its large-scale emergence can be prevented.

Warren Buffett has a lifetime of pattern recognition under his belt. The more attuned he is to these subjective pieces of contextual information, the better he gets at error reduction, just like cognitive security algorithms. Essentially, this cognitive pipeline implements the scientific method: it establishes and tests a hypothesis. In this way, it is not unlike Sherlock Holmes, literature's iconic detective. At the beginning of a story, for example, Holmes has no hypothetical scaffolding on which to hang his observations. Then, an initial piece of evidence raises his awareness. A second piece of evidence causes him to focus in. A third piece of evidence turns into three data points. Now he is working with a trend inviting further exploration. This is the frame. All of his activity now is about trying to fill the frame. "I think Ted murdered Mary with the candlestick in the conservatory," for example. Versus: "Mary was murdered and I have no idea who did it or how."

Cognitive pipelines leveraging multiple machine learning algorithms can do the exact same thing. An algorithm's investigation includes all of the following questions: Is this a benign request to me or is this somebody scanning for security vulnerabilities on my site? Is this just an end user or is it someone masquerading as an end user trying to hack my database? By using both structured and unstructured data—the subjective and contextual information that Warren Buffett exploits to his great advantage—the algorithm develops a hypothesis: *I think it is malware.* Now it is time to do some more research: to "read up" on pages and look more deeply at logs, network activity, or other elements of evidence that have relevance to the hypothesis. And after that process, the pipeline either yields a level of confidence that its original assertion is correct—yes, it is malware!—or achieves a level of confidence that this is not the case and no further research is required.

Science writer Maria Konnikova outlined the process of Sherlock Holmes in her book *Mastermind: How to Think Like Sher-*

lock Holmes. In it, she analyzes the thorny relationship humans have with their intuition in the midst of a pattern-recognition process.

> *Our intuition is shaped by context, and that context is deeply informed by the world we live in. It can thus serve as a blinder— or blind spot—of sorts. . . . With mindfulness, however, we can strive to find a balance between fact-checking our intuitions and remaining open-minded. We can then make our best judgments, with the information we have and no more, but with, as well, the understanding that time may change the shape and color of that information.*

Machine learning, however, does not have this dilemma when it arrives at its conclusions. When AI works, in my experience, the conclusions it comes up with or the actions it suggests are often *not intuitive actions*. If they were always intuitive, AI would be just the equivalent of a faster human researcher, but it is often quite a bit more.

This is a point I will revisit often in our journey together. Intuition—the bedrock of human intelligence—is that which is accessible and mentally convenient for us to chance upon. In complex problems, the utility function for a human—or the way we evaluate a series of choices in service of a goal—can be completely upstaged by AI. We will see examples of this in the financial markets later in chapter five. Even the proverbial "smartest guys in the room" are not smart enough to uncover counterintuitive patterns in data like AI can.

But before we move ahead, I need to share one more example of ransomware unfolding at the very moment I am writing this chapter: this dangerous new form of cryptoviral extortion sneaks in under a cloak of complete invisibility, roiling our globally connected systems and making the need for AI-based cybersecurity solutions even more urgent.

ADYLKUZZ:
DAMAGE ON THE DARKNET

While the security world was scrambling to address the damage done from WannaCry, a far more insidious virus started attacking systems and reaping the rewards on the Dark Web. In the spring of 2017, the hacker group Shadow Brokers made a number of Microsoft system vulnerabilities public. Although Microsoft quickly released patches for the flaws, many networked systems across the globe remained vulnerable. The Adylkuzz virus started taking advantage of these vulnerabilities, specifically through the NSA's leaked "EternalBlue" exploit. Adylkuzz does not spread at the same lightning speed as WannaCry, but it can enable even more nefarious ends. Instead of encrypting information, Adylkuzz installs hidden "miners" that generate a cryptocurrency called Monero. In essence, Adylkuzz does more than simply slow down your computer system or demand a payment for access. Instead it hijacks computer power to generate money for a gang of cyber criminals. The cryptocurrency it mines is similar to Bitcoin but is even more anonymous, making it a prime candidate for underground websites known to sell drugs, stolen credit cards, and counterfeit items. The longer Adylkuzz mines Monero on your system undetected, the more money is sent to your cyber attackers.

We need advanced forms of cognitive anti-malware programs to block this type of threat. By utilizing the power of machine learning algorithms, trained with millions of malicious files, AI security systems can—and did—identify the initial presence of this worm before its hidden miners were able to do much damage. Cybersecurity experts report that the most sophisticated AI solutions identified the presence of Adylkuzz as early as April 24, 2017, days before WannaCry was ever even announced. With the best of machine learning, Adylkuzz is shut down in a matter of seconds, not minutes. And all of this occurs before any human analyst ever enters the loop.

THE THIRD OFFSET

Of course, the military community is very attuned to these issues of safety and security. In January 2015, Deputy Secretary of Defense Robert Work outlined the Third US Offset Strategy to characterize our current national security environment. The first offset was nuclear weapons—or the application of massive amounts of force to deny the enemy an opportunity to attack. The second offset originated when nuclear weapons became too common and turned into a zero-sum game. That is when the defense conversation pivoted to small amounts of firepower applied extremely precisely. The apex of this second offset strategy was the Gulf War. But, inevitably, rogue forces and many adversarial nations now have access to these types of smaller, smarter weapons—similar to the US Tomahawk, a long-range, all-weather, subsonic cruise missile. Work argued that we have arrived at the third offset: supremacy in defense technology through rapid integration and massive application of artificial intelligence. This is autonomy at *every* level: distancing humans from many decision-action loops.

"In the 1950s and the 1970s," he stated in his White House briefing, "generally these advances were military capabilities that were brought along by military labs. But now with robotics, autonomous operating guidance and control systems, visualization, biotechnology, miniaturization, advanced computing and big data, and additive manufacturing like 3D printing, all those are being driven by the commercial sector."

In the age of the third offset, countries like China and Iran are trying to reassert themselves in their respective regions. They know their navies cannot match the conventional power of US carrier strike groups. Instead, they are employing asymmetric swarming tactics. China is also focusing on investments in AI-based cruise missiles.

The United States is responding turn for turn. In the fall of 2016, DARPA funded a project to deploy more than a hundred small drones on fast-flying F/A-18 jets. The fighters fly close to a target and release the Perdix drones, which then use autonomous algorithms to swarm around the target area. These experimental drones could be used for a variety of missions from surveillance to actual threat elimination. The old style of war—centralized command and control—is now being replaced by a new AI style of warfare. Along with it, the decision-action process—or the OODA loop (observe, orient, decide, act)—is rendering human intelligence less relevant. The speed of looming conflicts necessitates a new type of autonomous decision-making. We will explore this type of future war in extensive detail in the following chapter, "Warfare and AI," where it's AI against AI on the field of battle.

4.

WARFARE AND AI

Join me for a thought experiment originally published in the U.S. Naval Institute's *Proceedings* journal and conceived of by my friend and collaborator General John Allen of the United States Marine Corps, a four-star general and past deputy commander of US Central Command:

It is January 2, 2018, and a captain is contemplating damage to his ship after a surprise attack. This, however, was no ordinary attack. He is about to discover that this was a massive, widespread strategic surprise. Our captain and his crew had not anticipated the incoming swarm because neither he nor his ship recognized that their systems were under cyber attack. The undetected cyber activity not only compromised the sensors, but "locked out" defensive systems, leaving the ship almost entirely helpless. The kinetic strikes came in waves as a complex swarm of drones tore into the ship. It was attacked by a cloud of autonomous systems moving together with purpose, yet also reacting dynamically to one another and to the ship.

More than anything, the speed of the attack stunned and overwhelmed the sailors. Though the IT specialists on board the ship were able to release some defensive systems from the clutches of the cyber intrusion, the rest of the crew simply did not have enough decision-making time to react—mere seconds.

In these few seconds, some of the sailors ascertained, with their limited situational awareness, that the enemy's autonomous cyber and kinetic systems were collaborating. But in a matter of minutes, the entire attack was over.

The captain survived, and courageously remained on the bridge, but he was badly wounded, as was much of his crew. Fires were burning out of control and the ship was already listing badly from flooding. Because of the damage, the captain was unable to communicate with the damage control assistant who was herself badly wounded. It appeared that some of the autonomous platforms knew exactly where to strike the ship both to maximize the damage and reduce the chances of survivability. The captain's ability to command his ship was now badly compromised and the flooding was out of control.

After surveying the entire situation, he realizes he must make a call that no American skipper has made for generations. He issues the order to "abandon ship."

A TRANSFORMATION IN THE WAY WE WAGE WAR

Over the long arm of history, there have only been a handful of truly fundamental changes in the waging of war. Among these we can include the employment of cavalry, the advent of the rifled musket, and the combination of fast armor with air support and instantaneous radio communications in the execution of the Blitzkrieg strategy during World War II. Technological developments—sometimes originating from a variety of different fields—come together to enable these seismic shifts, and today we are on the cusp of another. Just as the Iraqi army floundered in the midst of the "second offset" technologies, including smart, precisely guided weapons, stealth, and electronic warfare, today's militaries must confront the fact that war is about to look radically different.

Broad contours of these changes can already be seen in today's warfare. Technologies like computer vision aided by machine learning algorithms, including deep learning, AI-powered autonomous decision-making, advanced sensors, miniaturized high-powered computing capacity deployed at the "edge," high-speed networks, offensive and defensive cyber capabilities, and a whole host of artificial intelligence–enabled techniques such as autonomous swarming and cognitive analysis of sensor data will be at the heart of this new revolution. All these capabilities coming together will augur something almost inconceivable: the minimization of the human in warfare. In this coming age, we will see humans providing broad, high-level inputs. Machines, meanwhile, will be left to do the planning, executing, and adapting to the reality of the mission while taking on the burden of thousands of individual decisions with no additional input. The human may eventually be left out of the decision loop.

HYPERWAR

General Allen was the first to call this AI-fueled, machine-waged conflict "hyperwar." During the Second World War, this term implied the global nature of the war involving many concurrent theaters. Today, however, what makes this new form of hyperwar unique is its unparalleled speed, made possible by automating decision-making and a concurrency of coexisting actions that will become more effective due to the leverage of AI and machine cognition.

Thus, in describing the wars of the future, military generals are using "hyper" in the original Greek sense of the word—"over" or "above"—a new type of combat that will be beyond what we've seen before in some very important ways. In military terms, hyperwar may be redefined as a type of conflict where human decision-making is almost entirely absent from the observe-orient-decide-act loop and, as a consequence, the time associated with an entire OODA cycle is reduced to near zero.

THE OODA LOOP

Up until the present age, the decision to act in warfare was always dependent on human cognition. But while human decision-making is tremendously potent, it also has its limitations regarding speed, attention, and diligence. As we have discussed earlier, there is a natural limit to how quickly human actors can arrive at a decision, and there is no avoiding the "cognitive burden" of making each grave decision. After a certain number, all human actors require rest and replenishment to restore higher cognitive faculties. Psychologist Daniel Kahneman reported on studies of this phenomenon with judges, showing that a simple factor—the lack of glucose, for example—can cause them to incorrectly adjudicate appeals. Tired brains slow down and even stop their analytical thinking, reverting instead to instinctive "fast thinking," which creates the potential for error. Machines, as we have discussed, do not suffer from these limitations. And to the extent that machine intelligence is embodied as easily replicated software, often running on inexpensive hardware, it can be deployed at scales sufficient to essentially enable an infinite supply of tactical, operational, and strategic decision-making.

AI ENABLES A DAVID AGAINST A GOLIATH

"Overpowering the enemy" is a phrase we often encounter in the literature of war. In military terms, this refers to the concentration of force in a finite space, over a finite period of time, such that the application of this force delivers a numeric or firepower advantage impossible for the opposition to counter or resist. This may not necessarily be because the attacking force is larger or more powerful than the entire defending force, only that *when and where it matters*, it is. This is a very important distinction. If a smaller force can be "perfectly coordinated" quickly and applied to a precise point that the enemy is unable to reinforce, then the smaller force will usually find success. If such action can be replicated repeatedly, then much larger opposing forces can be effectively neutralized economically and will often be dislocated psychologically. The superior force cannot achieve advantage and, what's more, the longer the encounter, the more rapidly that larger force deteriorates. This is the David and Goliath model of warfare.

The two key variables of concern here are time and space: the time required to form and execute an action of attack, and the space where such action is to be executed. Both these variables are computed as a result of significant strategic, operational, and tactical decision-making. Identifying the right spatial candidate for the application of force is the first ingredient. When done properly, this involves computing a large set of contingencies, called "branches" and "sequels" in military planning jargon.

With machine-based decision-making, a large group of sensors and shooters can be coordinated instantaneously, enabling the very rapid forming or massing of forces, and the execution of attack and subsequent scattering of the enemy. This machine-based decision-making is the principal fuel for hyperwar, and it will far outpace what can be done under human control and direction.

An old adage provides useful guidance for understanding how all warfare works: "Amateurs talk tactics. Professionals discuss logistics." Since time immemorial, waging war has required the movement of human armies that must be fed, clothed, and protected. The many and varied requirements of human soldiers drive the need for specialized roles that, in turn, require their own logistics. Force protection, medevac, and military policing only add to what is already a complex supply chain. In a fairly crude sense, a soldier is human intelligence + mobility + firepower. Modern armies certainly don't use human muscle as firepower any longer, and most of the time we don't employ human muscle for mobility, either. Instead, robotic soldiers will fill in these functions and they come in all shapes and sizes. Their needs will not be as varied as those of a human soldier, nor will they be as indispensable as a human soldier. Under almost all circumstances, a human commander would not consider putting lives at risk to extract a machine from the battlefield—a dangerous mission that may itself result in the loss of life.

Today's drones are mostly remotely piloted systems that simply separate the human pilot from the craft, placing human decision-making at a distance. This is a useful configuration but still has many downsides. First, the latencies involved mean that only certain types of missions can be fulfilled by today's drones—high-speed air-to-air combat would be difficult, for example. Second, the system remains susceptible to jamming and loss of communications. And last, the human pilot may succumb to many of the pressures and stresses of real war. The phenomenon of drone pilot PTSD (post-traumatic stress disorder) has been frequently documented, shedding light on the limitations of the current model.

Truly autonomous machines and robots in a variety of different types and sizes, with onboard synthetic intelligence, will be the foot soldiers in a future hyperwar. Models the size of commercial quadcopters, capable of weaving through forests and racing across open fields, will assemble, act, and disappear in no time. They will be armed with sophisticated sensors that feed vision and decision-making algorithms both on board, in the swarm, and, when accessible, in centralized locations. In addition to these, they will come equipped with a variety of cyber and kinetic payloads. A large number of these systems can be coordinated by means of swarm algorithms, enabling "a collective" to ensure the fulfillment of a mission and individual drones to support and to adapt to the loss of another.

AI SKILLS AND TRAINING

Erwin Rommel, the German World War II field marshal famously dubbed "The Desert Fox," once said, "The best form of welfare for the troops is first-rate training." Without training, there is nothing, and advanced forms of military training help create specializations for roles that are essential in the conduct of war. In the face of AI technologies and the hyperwar they will enable, we will see two groundbreaking changes in our approach to training. First, AI technologies such as natural language–based dialogue systems—capable of ingesting hundreds of thousands of pages of manuals, guides, studies, and more—will augment human operators in noncombat situations like equipment maintenance. Second, when employed in an entirely autonomous fashion, the tactics and strategies of AI can be easily copied from one system to another. This is the equivalent of having the most qualified veteran instantly communicate his or her experience and expertise to cadets who have never even participated in battle. Further, the skills and specializations they represent can be swapped in and out immediately. The same autonomous aerial platform can be an expert "pilot" for the suppression of enemy air defenses while a quick swapping of the neural network controller would make it the world's deadliest air-superiority specialist. What's more, if one such "expert" AI pilot needs to be sacrificed to achieve mission objectives, so be it. Other than the hardware, nothing is lost. After all, the AI "model"—the "brains" of the pilot—can simply be replicated onto a different piece of hardware.

Training for AI-based systems can happen in the real world, but it can also happen in simulators. Reinforcement learning—used in AlphaGo's crushing defeat over human masters in the ancient game of Go—is also being employed to build better and better autonomous cars. Each autonomous car doesn't have to go through the painful learning curve that every human driver must navigate. Instead, the car—or simulated car—that evolves the best-performing neural network can communicate its experience and learning to all other vehicles. This instant "transfer learning" will be another unparalleled reality in future hyperwar, fueled by the employment of AI.

A NEW AND BETTER THOUGHT EXPERIMENT

In light of these developments in the coming age of hyperwar, General Allen created another thought experiment to help us envision the future. Imagine it is now May 28, 2028, and our new ship's artificially intelligent cyber defense system just detected a possible cyber intrusion, maybe even an attack. The intrusion was so pervasive that it sought to "lock out" not just the ship's sensors and many of the ship's defensive systems, but also the ship's Anti-Swarm Batteries and supporting systems. The initial cyber attack and the successful defense occurred in less than a second. The defensive systems had functioned exactly as they'd been designed to do, so the ship was able to "sense," then detect, a massive incoming complex swarm attack—the kinetic follow-up to the invisible opening strike. And the system had gone even further, forwarding threat information to the rest of the fleet, enabling them to better prepare and defend themselves.

Our new captain moved quickly from the bridge to the main deck and, along with the others in the center, donned the augmented reality headgear and attendant gauntlets to assimilate and react to the totality and complexity of what he was about to lead. His first thought was his weapons status. He had only seconds now as some elements of the swarm were supersonic, maybe even hypersonic—or five times the speed of sound. Due to the elevated threat level, the captain had already been given a high level of autonomy to engage any potential attackers. He quickly cycled to the "weapons status view" in his headset, continuously fed targeting information from the ship's fire control complex, and all were ready. The captain had to act and so he shifted to the "status view." With a sweep of his hand into virtual reality, he initiated the view and, in that instant, made the one decision still available to humans in a hyperwar. He "decided" to open fire. Now "cleared hot"—or given permission to fire their weapons—all the various components of the ship's warfare system sprinted skyward outside the skin of the ship. The airspace was filled with several types of now completely autonomous aerial vehicles. Some moved off at high speed on the azimuth of the incoming attack to engage the enemy swarm at long range; others dwelt in the vicinity of the ship ready to engage as a last-ditch defense. No one on the ship, indeed no one in the US Navy, had ever experienced the ship's weapon systems going into action to full capacity. The structure shuddered and shook as each and every system leapt into the air with a cacophony no sailor in history had ever heard before. This was a moment as revolutionary as the Battle of Hampton Roads back in 1862, when Union and Confederate navies changed warfare forever with the use of ironclad warships, or, less than one hundred years later, when fleet action occurred outside visual range for the first time in the early carrier battles of World War II.

Back on board the ship, the captain shifted to "target view" in his headset to see what was coming in. He'd always been slightly skeptical of all these tools in his simulator training, but now he was seeing the reality of something nearly beyond belief. Completely autonomous aerial systems were locked in mortal combat: blue tracks representing his own systems, red tracks identifying enemy threats. As the battle unfolded—measured in mere seconds, and at incredible speeds—one after another red and blue systems winked out as they were destroyed crashing into each other or detonating in close proximity. The battle was moving toward his ship at a high rate of speed. In short order, the weapons officer spoke to the ship's AI through his own headset and unleashed the full might of the various close-in weapon systems, including the autonomous systems to engage the enemy swarm.

The first impact seemed deafening. Some elements of the enemy's swarm had detonated above the ship, taking out several of the "top hamper"—the ship's antennas. They were evidently "searching" for certain antennas to reduce the ship's connectivity. The second blast carried away the 20mm Phalanx Gatling gun, a principal means to defend the ship; the third struck the ship at the waterline, killing and wounding a number of crew and starting fires and flooding. While outside the ship a maelstrom was unfolding as kinetic systems autonomously coordinated fires with the near-continuous launching of the weapon system, inside the ship desperate damage control and medical recovery measures were now under way.

The captain quickly switched to "damage control view" and was able to see the AI-enabled damage dashboard as the ship's systems sought to fight fires and control flooding. Because of the sophistication of the AI system, he could instantly "see" which of the ship's systems were off-line, which were being rebooted to recover, and which were being instantly cross-connected to restore capacity and capability. The ship's system was autonomously shifting power loads, and bringing emergency systems online. Decisions for damage control were now made in seconds whereas dangerously long minutes were needed only a few years earlier.

Finally, the captain shifted to the view he dreaded: "crew status." Because every member of the ship's company was wearing a wireless "health status harness," measuring body temperature, heart rate, blood pressure, and breathing, he could instantly see the status of the crew overall, and each individual sailor's status dashboard. He dwelt here a few minutes, sobered and saddened by the number of casualties. As he cycled through subviews in this domain, he saw who had been killed and who was wounded. He knew which of his leaders were down, and he began to consider how he would reconstitute the chain of command. This was his family—his precious officers and sailors—and through this view, he knew where each member of his crew was located. He could see that the medical personnel, monitoring the same system, were deep in action as they already knew the extent of the casualties, the kinds of medical requirements now levied on them, and the locations where sailors were down or injured across the ship for recovery operations.

Hours later, with his wounded cared for, the fires out, and the flooding under control, the captain reflected on what had just happened. He was shaken, but not frightened by the reality. The attack had come seemingly from nowhere. The cyber defense system had detected the initial cyber intrusion, and had not only protected the ship, it had reasoned the extent of the attack was a precursor to something larger and alerted the ship's chain of command of what might be coming. This hypothesis had been formed, researched, and validated in less than a second. Within ten seconds, the ship initiated general quarters on its own, and the captain had already donned his augmented reality ensemble. From that moment until the final fires were put out was approximately two or three minutes. The autonomous nature of the weapon systems and the ship's defensive systems had foiled a coordinated, complex cyber and autonomous swarm attack. The captain was struck by the realization that at nearly every point where human actions and decisions were required, the fate of the entire ship was at a far greater degree of risk. He had just experienced the near mind-numbing speeds of AI and deep learning–driven warfare. Indeed, he suddenly realized he'd been the first American commander to fight in the environment of hyperwar. He certainly would not be the last.

A MILITARY REVOLUTION

This scenario presents just a few of the ways in which synthetic intelligence will fuel the next great shift in how warfare will be conducted. The fusion of distributed machine intelligence with highly mobile platforms brings an unprecedented speed and scale to military action. Advances in AI have the capability to fundamentally change the human condition, and with it a profoundly human undertaking: war.

Near-peer opponents are already investing heavily in these technologies and have even operationalized some AI-powered weapon systems, such as cruise missiles. Autonomous algorithms can now transform moderately dangerous weapon systems into threats impossible to ignore.

As our thought experiment illustrates, the speed of battle at the tactical end of the warfare spectrum will accelerate enormously, collapsing the decision-action cycle to fractions of a second and giving the decisive edge to the side with the more autonomous decision-action concurrency. At the operational level, commanders will be able to "sense," "see," and engage enemy formations far more quickly by applying AI algorithms.

This will result in swarms of complex, autonomous systems capable of simultaneously attacking the enemy throughout its operational depth.

And at the strategic level, the commander supported by this capacity will "see" the strategic environment through sensors active across the entire theater. AI-powered assistive technologies such as intelligent assistants, advanced interactive visualizations, virtual reality technologies, and real-time displays projecting rapidly updated maps will all come together to enable instantaneous situational awareness.

All of this reawakens the perennial conversation about the nature of war. If we are poised at the edge of something called hyperwar, we must explore adapting to this new conflict environment, especially to the moral dimensions. Since the prospect of full machine autonomy—overall range of action, including deadly response—is disconcerting to many, public debate on this topic is infused with softeners. These are comforting terms such as "semiautonomous" and "human in the loop." However, these represent an easy out while also being misleading. They masquerade as answers when they don't even begin to address the questions. Effective machine functionality in a variety of situations requires full autonomy, and a wink and a nod to a "man in the loop" is actually detrimental to properly confronting and addressing this need. For example, how do we expect a swarm of autonomous undersea vehicles to act when they have a critical target in sight—say a North Korean submarine about to launch a nuclear-armed ballistic missile—but realize that communications are being jammed? Do they let the threat materialize since they can't contact their human commanders? Or do they take autonomous action for our protection?

As a further extension of these types of questions, we also

need to understand that we are already living in an age where the only viable defense to certain types of smart missiles comes from AI. For example, US naval ships currently include equipment like the Phalanx CIWS (close-in weapon system) with a radar and a computer mated to a Gatling cannon that generates a very high rate of fire. Once engaged, it is a completely autonomous system designed to protect ships against sea-skimming cruise missiles and other aerial threats. When a missile enters a two-and-a-half-mile radius around a ship, a human doesn't have enough time to react. The Phalanx system *must* operate in a completely autonomous way. It tracks the missile, aims, and fires completely on its own.

Hyperwar will also change our understanding of aerial combat. In the old dogfights of the mid-twentieth century, the pilot's skills were the difference between life and death. Fighter pilots had to get the higher ground and higher energy position if they hoped to fire because recoil and the energy advantage of the target would make it impossible to successfully engage from below. In the '80s and '90s, however, pilot training started to focus mainly on BVR technology—beyond visual range. Instead of one pilot in a plane seeing another pilot in a plane and engaging in a dogfight, the radar technology now detects the target at long range and the pilot simply shoots the missile. The earliest models of BVR missiles were not very accurate due to the lack of sophistication in both sensors and algorithms, but these capabilities have evolved significantly today. The latest missile guidance systems provide true fire-and-forget capability. A missile such as the American AMRAAM (advanced medium-range air-to-air missile) can hit its target without the launcher even being in the line of sight of its target. In essence, the fighter aircraft is already turning around and disappearing before its missile has even approached the other plane. There is a reason they call it "fire-and-forget": the weapon is doing the thinking, the pilot is the one forgetting.

There are, of course, low-tech threats in global warfare as well, but even these require more sophisticated automated targeting systems. In 2012, the Iranians wanted to warn the US Navy not to send its aircraft carriers into the Persian Gulf. They showcased hundreds of their hydrofoil boats that skim above the water to achieve extremely high speeds. Iran's implicit threat was that they could put five hundred or even a thousand of their men in these "go-fast boats" and send them out in a "swarm" to overwhelm US carriers. It is a strategy not unlike that of the Japanese kamikaze pilots in World War II because the hydrofoils would essentially become suicide boats. Such a strategy would be unlikely to stop a carrier altogether, but the technique could certainly take a toll in terms of lives lost.

As with all of these examples in our future military landscape, there is no genuinely effective counter to this type of swarm technique aside from AI. It is simply not practical to place dozens of gunners on the deck of an aircraft carrier. Even if it were a practical possibility, human gunners would not be able to respond quickly enough to the swarm of boats and their launched projectiles. The only viable response to a swarm is a counterswarm, or an automated targeting system controlling a weapon such as the US Navy's LaWS (Laser Weapon System).

It goes without saying that we are not alone in this drive toward hyperwar. In 2016, for example, Wang Changqing of the China Aerospace Science and Industry Corporation told the state-run *China Daily* newspaper that China was developing an AI-based cruise missile. "We plan to adopt a 'plug and play' approach in the development of new cruise missiles, which will enable our military commanders to tailor-make missiles in accordance with combat conditions. Moreover, our future cruise missiles will have a very high level of artificial intelligence and automation," he added. "They will allow commanders to control them in a real-time manner, or to use a fire-and-forget mode, or even to add more tasks to in-flight missiles."

The message of such developments in military capabilities is clear: we need to train and educate our leaders—young and old—to think in different ways and be comfortable in these upcoming environments of fast-moving decision-making. With the ANI developments on the near horizon, it will almost certainly work to our disadvantage to insist on putting human intelligence in the loop at all times. As we will discuss in the chapters to come, ANI developments like an AI-based cruise missile inexorably lead to a future where human input in certain conflicts is not only unnecessary but also dangerous.

If our human input is no longer needed in certain aspects of actual warfare, how can we, as citizens, remain engaged in the development of our military? To answer this vital question, I reached out to another of my esteemed colleagues in the military, Lieutenant General Ken Minihan, a retired Air Force officer and former director of the National Security Agency (NSA). General Minihan has had a long and storied career in the military and government for close to fifty years; while serving in the Clinton administration, he was one of the first advisers to publicly recognize the unique threat of security in the twenty-first century.

"Twenty years ago," he told me, "no one was really presenting a coherent discussion of the vulnerabilities of the new technologies. You could see, in the transition to a post–Cold War, that the government's responsibility for security with these new threats was being abandoned in a budgetary context. We were developing all these new technologies commercially but we weren't developing any way to make them secure or private. My motivation was to start to tell the story that transitioned the responsibility for privacy and security from the government to industry. And to try to create conditions that made that private sector feel accountable."

According to General Minihan, we are still navigating this transition of accountability in the cyber age. Unfortunately, the threats have only increased. In 2015, a secret NSA map obtained by the press showed that the Chinese government orchestrated close to seven hundred corporate, private, or government cyber assaults. Their most recent military publication, *The Science of Military Strategy*, put out by top research institutes of the People's Liberation Army, explicitly acknowledges that the country has specialized units for waging war on computer networks. And as we all know from recent news briefings, China is far from the only country developing increasingly sophisticated cyber warriors. Russia, Iran, and North Korea—not to mention terrorist groups—are all ramping up their cyber warfare forces.

However, this issue of accountability—General Minihan's goal of making the private sector take on more responsibility for security concerns—is often complicated by questions regarding privacy. In a democratic system like the United States, people are decrying the dangers of a police state: a dystopian *1984* where the government is watching everything we do. Although I appreciate bringing all these concerns to the public square— debate is necessary for our democracy to thrive—I also lament the at times crippling nature of this particular discourse. Here in democratic countries, we don't function under a police state.

But what about the *real* police states that are already using advanced technology and will continue to use it to further their ends? While affording protections to citizens is paramount, should we curb our government's ability to mount a credible defense against foreign adversaries?

We saw a similar curb placed on genetics and stem cell research during the Bush presidency. Those same impediments did not exist in China, and in 2010, the Chinese created the largest next-generation sequencing center in the world in Shenzhen. The technological differentiation that we have always prided ourselves on is slipping away. And all because we insist on seeing the arguments in binary terms: security or privacy; artificial intelligence or humans; utopian immortality or the dystopian nadir.

"I see security and privacy more as the two rails on a railroad track," General Minihan told me. "They go up and down but they stay symmetrically associated with one another. You don't have one or the other; you have a balance that can be shifted based upon threats and new technologies. That is why I prefer to have a conversation around 'Can I trust this technology?' not 'Is it secure?' or 'Is it private?' "

Here, I argue, is where smarter software and artificial intelligence can help form a more nuanced third way. For example, in 2013, there was an outcry from privacy advocates about the Transportation Security Administration's (TSA) use of "back-scatter" body scanners in airport security. They were vilified as "virtual strip searches" and eventually TSA removed all of the machines from active use. For much of the public, the violation of privacy occurs at the point where the information is shared with another human for interpretation.

Just as artificial intelligence can advance so many of the scientific endeavors addressed throughout this book, it might also mitigate these privacy concerns. We can use body-scanning machines in concert with cognitive software that generates an automated analysis of the captured imagery. Any images would only ever be shown to a human if the software detected an anomaly with a high degree of confidence. This could limit the exposure of private citizens to oversight by the government, the source of much of the problem. AI and smarter software can be used to strike a balance of *trust* between privacy and security. This *trust component* is a vital part of my exhortation for a more nuanced discussion: a third way. Developing better software and smarter analytics is not going to solve all of our security problems, nor is it going to be the road to our demise. The reality lies somewhere between the two: yes, the risks from technology can and should be managed, but it just so happens that the best way of doing this is by developing more technology—specifically more *trustworthy* technology.

Let's revisit the current ban against autonomous weapons proposed by Musk, Hawking, and other heavyweights as an example. This proposal cites ethnic cleansing as one of the most dangerous applications of autonomous weapons, beyond bioweapons that target certain types of DNA. For example, an autonomous weapon system could be programmed to seek out members of a particular racial group and then eliminate them. It's true that at least a crude version of such a drone is very much within the realm of what is commercially available today; it might even be obtained at a hobby-level price of only a few thousand dollars. If it were coupled with autonomous guidance software sourced from a community like DIY Drones—an online discussion forum for developing control systems founded by former *Wired* editor Chris Anderson—such a system could easily be created. This is not the AI of tomorrow; this is the possible AI of today. When the drone happened to see an individual of a particular color, complexion, or with certain physical features that indicated their ethnic background—using free and openly available image recognition software—the drone could then engage the identified individual as a target. The whole endeavor could be developed in a small private facility with a limited budget in a matter of weeks.

Considering all this, one would think that I would be on the side of Musk and Hawking. After all, this is a viable threat. And yet it is the very fact that this technology is so readily available that leads me to take issue with their efforts. How could such a ban ever be enforced? Is a statement by former secretary-general of the United Nations Ban Ki-moon going to stop North Korea from developing autonomous weapons? Or China from shipping CSS-2 missiles to unsavory governments in the Middle East? Or ISIS? It is simply not practical to expect that an agency or an international treaty will effectively monitor such activity. I assert, once again, that the AI genie of innovation is out of the bottle; it cannot be stuffed back inside.

Before we parted ways during our most recent conversation, General Minihan left me with a provocative thought. "When I talk to today's generation of leaders," he said, "I always say, 'How much do *you* want to spend on intelligence? How much do *you* want to spend on the military? How much do *you* want to worry about a cyber attack?' I know the answers to all those questions in the industrial age. But these are not my questions to answer in the cyber age. This generation has got to start speaking up and come to some conclusions, just like your grandparents did after World War I and World War II. Here are the lanes in the road: What is the strategy? It's up to your generation to say how you want to take care of our world in the twenty-first century."

I have an answer to General Minihan's challenge. I want my generation to be prepared, and preparation means developing technology to manage these viable threats. We can't depend on more unenforceable bans and treaties. We need next-generation AI-based technology. And we need to foment societal discourse that leads to a balanced view of artificial intelligence.

Fortunately, for those in the military as well as for all citizens, a new petri dish for technological progress and AI-embraced innovation is happening outside the corridors of computer science departments and the start-ups of technology hot spots like Austin, Texas, and Silicon Valley. Today, as we will see in the following chapter, some of the most cutting-edge AI algorithms are being trained in a low-friction system with instant feedback mechanisms: the financial markets.

5.

FINANCIAL MARKETS

In 1915, an extraordinarily clever young economist and former civil servant took up an official position at the British Treasury. Among other things, he was tasked with acquiring the scarcer foreign currencies to purchase war supplies. The deft young trader started to hoard the Spanish pesetas, removing them from the marketplace and, as a result, increasing their value. When his superiors in the British government told him that they urgently needed Spanish pesetas to pay off their debts, our trader took all of his hoarded currency and flooded the market with them, quickly reducing their worth. He then immediately turned around and bought them all back at the now significantly cheaper price. Our young speculator walked away with enough money for Britain to pay off its debts, pocketing a 30 percent profit in the earnings. His colleagues later described his derring-do in notes and journals with reverence and awe. He was the first trader known to manipulate the global currency market. His name was John Maynard Keynes.

Today, Keynes is widely celebrated as the father of macroeconomic policy, but a lesser-known fact is that he was also a stunningly successful financial adviser and fund manager. During the two decades that he managed the trust of Cambridge University, he garnered a return of 400 percent—all of this during a period when the overall stock market in Great Britain changed very little.

One of Keynes's greatest insights was his ability to accurately observe the difference between a human estimate of likelihood and a formal mathematically driven probability assessment. In essence, he identified the fact that the value of a stock is merely what any one group or another is willing—irrationally—to pay for it. He used this understanding to exploit the herd mentality in the marketplace and he died with a net worth of $30 million, money made almost entirely from his own investments.

Extraordinary investors and market theorists all have one thing in common: the ability to act in a contrarian fashion. Warren Buffett warned, "You pay a very high price in the stock market for a cheery consensus," and Baron Rothschild, member of the famed Rothschild banking family, famously advised, "Buy when there's blood in the streets, even if the blood is your own." When a great investor discovers an alternative way of thinking about the markets, the investment is rewarded. Great fortunes are almost always made at the hand of a counterintuitive approach.

Because they rely on maverick personality types—nonconformist to an extreme—we rarely encounter these masters of the market in everyday life. Once in a decade or two, an investor emerges from this culture with the ability to come up with insights that the rest of the world simply cannot see. This is why hedge fund managers like Alan Howard and George Soros deserve their starring roles in history. But as the pace of trading speeds up and speculative insights are perceived, not in days or hours but in *fractions of a second*, a new leading character is poised to break out onto the stage. Enter AI . . .

If you pull up the list of available jobs at hedge fund Two Sigma Investments, based in New York City, many of the most senior-level positions have more to do with computer programming than they do with investment knowledge or finance. Instead of waiting another ten or twenty years for the next market wunderkind to appear, hedge funds like Two Sigma are predicated on an entirely different strategy. Whereas the discretionary hedge funds of the 1980s and 1990s—run by big-name players like George Soros, Paul Tudor Jones, Bruce Kovner, and the father of them all, A. W. Jones—were based on human expertise, a fund like Two Sigma has little human input interfering with a trade. Instead, a mathematical idea is mechanized and automated to run the entire trading strategy. When Two Sigma hires new traders, they are not looking for the stock trader archetype of yore, the alpha male who makes big bets based on personal experience with market patterns. Instead, more and more, these systematic hedge funds are being infused with AI: many of their "traders" are actually mathematicians and computer scientists in disguise. In the past, a trader like Keynes could creatively devise a strategy to manipulate the price of the peseta and walk away with some tidy returns. Today, however, with increasing speed and transparency, everyone around the globe can see everything, including any and all speculative attacks on a currency. Hedge fund managers are struggling to come up with unexplored moneymaking strategies. The market is tapped out for true financial innovation. Or, at least, that is how it appears to us.

• • •

When AI is working at its best, its choices look like sheer madness. While humans are limited to the world of reason and knowledge gained through our bodies, AI and its purpose-built mind is free to roam. And roam it does. Because AI is not constrained by any physicality, its travels take it to the realm of the ridiculous where it mines for brilliant breakout strategies.

Humans have no accurate parallel to this experience. We can never really understand that, beyond the constraints of our own human rationality, there is a thrilling bounty of creativity. In essence, AI can be mechanized madness and yet, at the same time, its intention is more rational than we can ever hope to be.

Allow me to illustrate this point with a brief thought experiment: take two minutes and think about the name Led Zeppelin. When you read these two words, the name of the iconic rock band, chances are your mind was filled with a variety of associations. Led Zeppelin means several different things to you all at the same time. When I hear the name Led Zeppelin, for example, the atmosphere of the library on the University of Texas campus is conjured up in my mind. Suddenly, I am thinking not only of Led Zeppelin and the library on campus but also of my best friend, Zaib, who I am now lucky enough to call my wife. From there, the links only continue and the single thought of Led Zeppelin lights up the associations across my consciousness: the links of my years as a university student, my exposure to the computer science department at UT, the way the music created the shape of so many of my evenings with longtime friends from that period in my life.

This—my Led Zeppelin example—is the way humans think.

Our sensations are linked together and we find context in this: our associative memory.

I cannot delink the notion of Led Zeppelin from all those other experiences in my mind. It anchors me in a time and place: it is the centroid around which I can travel yet never entirely escape. Because of this, I will always experience Led Zeppelin in a certain way. Of course, over the years, other people have exposed me to their associations of Led Zeppelin and my understanding of the band has expanded to include a wider variety of anchors. I now experience Led Zeppelin in more ways but I will never experience Led Zeppelin in *all* ways. I can never achieve what the philosopher Thomas Nagel described as "the view from nowhere." I will never have an entirely objective perspective on Led Zeppelin.

These associative memories work across multiple people, across entire cultures. When you introduce a concept, you bring your ideas around that concept and you plant those in my mind. Some of the associations I make on my own, but I am never far from your centroid on the subject. These become the common association of a great many people. This is culture: it is an overlapping tapestry. It is very difficult for you to introduce Led Zeppelin to me in a way that is completely different from your associative context of this name. In this way, we all share not just our associations of a concept but also the associations *between us*. We can call this intersubjectivity. To see us all as individuals and to see all thoughts as occurring in atomized bubbles is misleading. Thought is a societal, not an individual, endeavor.

Where, for example, do our most innovative ideas come from? Given the nature of this associative memory, they come from years of accumulated learning, experience, experimentation, and, on occasion, flashes of inspiration. Our humanity—our interlinked minds—form the ideascape of shared associations. So much so that cutting-edge ideas can be discovered by people at the same time living across the globe from one another: famed physicist Steven Weinberg had a flash of insight linking electromagnetism and the weak interaction that controls nuclear decay. Physicist Abdus Salam of Pakistan was circling around the exact same idea at the exact same time. In 1979, the two scientists shared the Nobel Prize in Physics, along with Sheldon Glashow.

This is the work of our associative memory. But as we have discussed earlier in our journey, humans don't have purpose-built minds. They have general-purpose minds that allow them to exist and thrive in the physical world. So while experts like Weinberg, Salam, and Glashow have innovated at the edges and borders of human knowledge and common sense, they are inherently bound by their own biological limitations and associative memories. Their grounding in the physical world—the anchors that keep them tethered to our collective societal notions of common sense and intelligence—are the very things that impede them, and us, from coming up with ideas that are beyond human skill or imagination, beyond the borders of human knowledge.

With AI, our intellectual ideascape becomes limitless. Our associative memories and pruning have aided us greatly over the longest arm of evolution because we have been able to efficiently get rid of any ideas deemed "unnecessary." Anything counterintuitive, nonsensical, or considered downright "mad" immediately gets dismissed and most of it is never dreamed up. It is delinked from the dense web of associations we carry among ourselves.

And yet, when it comes to a frictionless landscape like the financial markets, rationality under the guise of nonsensical madness is exactly what is required. We can see how this works in the human responses to a particularly "mad" move made by AlphaGo—the reinforcement learning algorithm that recently defeated the best Go player in the world, Lee Sedol. Fan Hui, a champion Go player, saw the move and was astounded by it. "I've never seen a human play this move," Hui told reporters. *So beautiful.* Sedol, for his part, was so unnerved by the move that he had to get up and call a fifteen-minute break after taking it in. He told reporters later that he was sure it was a mistake. As *Wired* magazine reported in its coverage of the competition:

> *AlphaGo's move didn't seem to connect with what had come before. In essence, the machine was abandoning a group of stones on the lower half of the board to make a play in a different area. AlphaGo placed its black stone just beneath a single white stone played earlier by Lee Sedol, and though the move may have made sense in another situation, it was completely unexpected in that particular place at that particular time—a surprise all the more remarkable when you consider that people have been playing Go for more than 2,500 years. The commentators couldn't even begin to evaluate the merits of the move.*

So what does this beautiful madness look like when it is applied to the financial markets? The systematic hedge funds of the last decade have certainly achieved success, but their pre-set formulas were unable to learn and adapt. When the reality of a volatile market fell outside the bounds of their models, their formulaic intelligence inevitably stalled. With the advent of AI, however, and algorithms such as deep learning, models are able to look at vast amounts of market data and adapt their strategy to exploit market movements in real time. This cadre of silicon-based traders has no need to drink, rest, or eat, no desire to wind down with a beer or sleep in on Sunday. Companies like Goldman Sachs have a punishing work expectation for their newest recruits. Young hires straight out of college often spend nights under their desks at the legendary firm, sleeping on the famously dingy carpets and eating cold pizza for breakfast. Reinforcement learning and other AI algorithms are now giving all this young talent a run for their money. These algorithms discover new strategies and recognize new patterns at a rate faster than any human analyst could. And once they find a promising hypothesis, they can usually find many related strategies even faster. The rate of this discovery can be exponential. Whereas it takes years of experience as a "market maker," a trader, to train the human brain sufficiently to recognize patterns and intuit unique strategies, algorithms are currently at work uncovering thousands of workable strategies within hours. How's that for coming up to speed?

Unlike opportunities in real estate, say, or manufacturing—industries that exist in a friction-filled world of physical objects—the financial markets are one of our society's *least* friction-filled domains. In the markets, it is possible to generate a hypothesis, test it, get immediate feedback, and then either continue with it or pivot on to something new, all in the time it might take for a real estate mogul to unfold a blueprint. Very few domains exist so completely in the abstract, represented entirely by numbers, and with such a tight feedback loop between action and consequence. While a master like Keynes is famous for innovating at the edges of human knowledge and common sense, the new AI-augmented algorithms developed by a hedge fund firm like Two Sigma allow traders to surf the landscape of ideas at speeds previously unfathomable. Deep networks and other sophisticated digital structures generate hypotheses that go beyond what unaided humans can comprehend. Madness with a method ensues.

And hedge fund managers are ready to go mad. Over the last decade, in the post-crisis environment, discretionary hedge funds in particular have reported abysmal returns. Since 2009, hedge funds in general have underperformed the S&P 500 by 51 percentage points. In 2016 alone, discretionary superhero Paul Tudor Jones fired 15 percent of his workforce and announced that he was moving his firm, Tudor Investment Corp., to a more systematic, computer-based quantitative fund. Louis Bacon of Moore Capital Management, who once took in returns of 15 to 20 percent, was down to 5 percent in 2016. Bill Ackman of Pershing Square Capital Management had a disastrous 2015 when his fund lost 40 percent. Experts estimated that Pershing Square Capital Management would need to rally by more than 60 percent in the following quarters to make up for the loss.

More and more, discretionary managers are hopping over the fence to join the "quants." A 2016 Goldman Sachs study reported that one in five US-based investors were investigating quant strategies. Just one year earlier, in 2015, that number was only one in ten. AI algorithms, combining deep learning, deep hashing, natural language processing, reinforcement learning, and genetic techniques, can augment more traditional hedge funds as they uncover counterintuitive trading strategies, predict volatility, and identify patterns and event correlations that would never be accessible to the human eye.

• • •

The copy of Sun Tzu's *The Art of War* I once owned in college included a famous story in the forward. A great Chinese general was confronted in a war with an army and general far superior to his own. He set out a command to his five thousand troops. When they confronted the enemy, the smaller army formed a single line, shoulder to shoulder. Then, upon hearing their general's command, they all took out their swords and slit their own throats simultaneously. The opposing army first stood in horror at the display. Then, one by one, they ran away in fear.

Sun Tzu tells us that the best generals win the war before it even begins. They strike fear in the hearts of their enemy not through brute force but with a tireless and obsessive commitment to the goal and to the empire.

AI offers us a vision of Sun Tzu's perfect mythical army. This is an army endowed with compulsion, an army that lacks all fear. This is an army that will stay its course, no matter its adversary.

In the face of stunning market free falls after, say, the collapse of Lehman Brothers on September 15, 2008, can a hedge fund manager say the same? During the Great Recession, how many market warriors stayed the course with their swords in hand? Like the Chinese army of yore, you simply can't scare an algorithm. And fear is what impedes most traders from making the most exciting counterintuitive moves in the market.

In 1996, Garry Kasparov was revered the world over for his mastery of chess. But he was also known for his psychological warfare during competition. It was chalked up to the advantage of the "stare," that incalculable extra touch of intimidation Kasparov used to knock his opponents off kilter. Just as he dominated the board, he also dominated the psychological space in the room by locking his opponents in with his eyes.

When he played against Deep Blue, IBM's supercomputer,

however, Kasparov came face-to-face with Sun Tzu's army, cast in silicon and steel. Kasparov stared long into Deep Blue's eyes but there was only will, intent, and determination there to respond. AI holds a mirror up to humanity and Kasparov found himself staring at his own reflection.

AI AND BLOCKCHAIN: THE MATHEMATICS OF TRUST

Just as AI is disrupting the entire hedge fund industry, it is also transforming our cultural conception of the "financial institution." Throughout history, financial institutions such as banks, indexes, and clearinghouses have been providers of trust and low-friction transactions. Ever since we entered the colonial era of empires, joint-stock companies, conglomerates, and global finance, these institutions have served as a trustworthy intermediary between you and me: I owe you money so I give you a bank draft that feels as good as cash. In reality, the draft is nothing like cash, but in our collective imaginations, the story instills confidence. We trust in the ability of the financial institution to pay the money. The more we reduce friction—bank drafts are decidedly easier to carry around than shells, a herd of oxen, or crates of tea—the more we can trade. And the more we trade, the more our belief in this narrative is affirmed. When we make an exchange of goods, you don't really "trust" me—a complete stranger living halfway around the world—you "trust" in the bank. The conceit of the "bank" makes manifest our faith in cooperation en masse.

Until very recently, "bank" meant an actual building. We might even call it a "branch." This was a physical place where we did our "banking." Inside, we often brought our checks to a human teller and passed them over to be processed, placed in a clearinghouse, and then, days later, transformed into numbers on our balance ledgers. Many of us still remember the bankbooks they used to pass out to new customers. They came wrapped tight in a plastic sleeve, each transaction recorded under the watchful eye of a teller. Bank branches, for many of us, still conjure up images of stately wooden desks, leather couches, and pens attached with long strings of metal beads.

All of that, today, is quickly becoming obsolete. Even the idea of the financial institutions—the story we all agree on—is up for debate. This is due, in part, to an innovation created by a mysterious cryptographer who goes by the name of Satoshi Nakamoto.

In 2014, *Newsweek* reporter Leah McGrath Goodman showed up on the doorstep of Japanese-born engineer Dorian Nakamoto and "outed" him as the elusive creator of Bitcoin, the digital peer-to-peer currency designed in 2008. The article immediately created a buzz despite Nakamoto's vehement denials of any involvement in the cryptographic endeavor. Although *Newsweek* stood by Goodman's claims, more and more sources close to the Bitcoin project started to doubt that Dorian Nakamoto was the originator. Who is the real founder of Bitcoin? As of this writing, we still don't know, although a new goose chase is on with Australian cryptographer and businessman Craig Steven Wright. He seems to have the technical knowledge required for the development of Bitcoin, but does he have the singularity of mind, the "genius," to create a code that cryptographers around the world deem downright brilliant? It is, so far, an impenetrable mystery worthy of the man—or woman—who has likely changed our entire understanding of cooperation and risk. What we do know is that Bitcoin, and more specifically, blockchain, the underlying storage system upon which Bitcoin transactions are recorded, is rapidly upending major aspects of the financial system.

Let's pause for a brief primer on what these digital entities actually are. Bitcoin is a digital currency that was originally designed to be a distributed, peer-to-peer (P2P) method of payment that could function in the absence of any central authority governing its use. We know about peer-to-peer distribution from such infamous examples as Napster and the PirateBay, but P2P is about far more than downloading hard-to-track music. It is fundamentally about the resilience gained from eliminating a single point of failure. At the heart of Bitcoin and many other similar cryptocurrencies such as Litecoin and Dogecoin is a secure, distributed, peer-to-peer database called blockchain. Like past P2P services, blockchain is a storage system capable of preserving digital information and its provenance—the complete history of how every piece of information was added, modified, or removed. Of course, while cryptocurrencies like Bitcoin use a blockchain, this technology can be used for much more. In fact, the information stored in a blockchain can be anything from addresses to land deeds, stock trades, and patent and intellectual property records. The guaranteed security provided by blockchain ensures that no bad actor, or even a large group of bad actors, can corrupt or malevolently modify its records. This inherent resilience enabled by the blockchain allows its use in Bitcoin and other cryptocurrencies as the trusted ledger for tracking currency ownership and transfers. While traditional currencies depend on a bank as an issuer, repository, and validator, with the blockchain there is no need for such a central authority. No bank, no branch, and no need for authentication. The math behind the algorithm ensures security and trust. In fact, the blockchain is so resistant to improper manipulation that any information in this distributed data store is guaranteed to be accurate in all circumstances unless more than half of the participants in the blockchain have been compromised. With ten or twenty participants, a security breach might pose a realistic risk. But at this point in the adoption of blockchain technology there are millions of users on any one blockchain or another. The highly unlikely possibility that half of those millions will all become compromised makes the blockchain one of the world's most widely used and secure storage systems for important information exchange.

To better understand just how revolutionary the blockchain can be, consider how it might upend our understanding of a financial institution. Let's take an example from our everyday interactions with banking. Today I go and deposit a check with my bank. The financial institution says, "Okay, you have received $100. We will add this to your account. So instead of $500, now you have $600." Going forward, anyone who wants to receive money from me has to rely on my bank as the only central repository of information concerning my account. In this way, the bank becomes an empowered intermediary as the only keeper of my account information. We might even say that banks are not simply in the lending business; they participate equally in the *information* business.

When a financial institution transfers money to somebody, they don't actually transfer physical *currency*. They just make a record that says, "Bank of America owes JPMorgan Chase two million dollars based on the transactions that happened today." Then they use a clearinghouse to settle their actual records and manage the risk as the "money" is being moved from one account to another. For the most part, transactions are really just electronic records: information. And we rely on financial institutions for the veracity of this information.

With blockchain, on the other hand, all of these steps are obviated. Let's look at what happens when a transaction occurs between people using Bitcoin. Say you want to pay me ten Bitcoins from your account. Instead of writing a check to some third party, you simply "write" the details of the transfer to the blockchain. That means *every* participant in the blockchain sees that transaction. Your anonymous ID sends ten Bitcoins to my anonymous ID and the transfer record is replicated all over the chain. Within eight to ten minutes, everyone on the chain is able to ascertain whether this was a legitimate exchange. If the network disagrees with what was initially received, participants can discard the "bad" block, or the "bad" chain of blocks.

What this means is that all the participants get to see every transaction by every other participant, and encryption ensures no block can be forged.

In its essence, blockchain is a disintermediation of trust with the use of mathematical guarantees. Our subjective notion of trust—based on religion, art, and poetry—can now be expressed as a computationally guaranteed property of a system.

In the age of AI, where notions of algorithm-enforced safety, decision explainability, and transparency of action will become key concerns, the blockchain provides one example of how a human ideal—i.e., trust—can be translated into mathematics and code. In fact, work I am presently involved with builds on the blockchain to allow multiple AI agents to develop a shared, incorruptible view of their world. If they perceive the presence of AI actors—even agents from within their own collective—that are "breaking the rules" or have gone rogue, the rest of the collective negotiates on and agrees to a response. This might be the first time a blockchain is being used to enable the idea of "social responsibility" in AI agents. Our progress thus far shows great potential.

No one is thinking more about this redefined trust in financial markets than Chris Corrado, COO and CIO of the London Stock Exchange Group (LSEG). Corrado has worked in finance and technology all over the world, and though he is a firm believer in increasing automation across all aspects of the financial industry, he contends that risk management will never be completely "robo-sourced."

"One of the most critical things we do," Corrado told me in conversation, "is clearing trades." As a clearinghouse for global trades, the LSEG is responsible for managing enormous amounts of risk on behalf of its customers in real time. "The blockchain reduces the lag time between knowing that a transaction is completing and ensuring that the money is moving to the right place. Anytime you can reduce that inherent risk, you are making a move that is better for society. You can employ that capital to better use."

Yet despite the promises of the technology, Corrado remains convinced that there is still a place for trusted institutions. "As your role evolves, the activities that you do change," he told me. "This is not because the activities are unnecessary, but because you are using an enabling technology to do them. And as a result, you need fewer people to execute those activities."

Imagine for a moment you have to wait until the end of the day to find out how things get reconciled versus finding out in real time. The intraday risk that you are taking is materially different. That doesn't mean it doesn't exist. You can do snapshots to make it less severe, but if you're not guaranteeing on a transaction-by-transaction basis that the money is going to end up in the right place at the right time, then, by definition, you take risk.

This is the future. Essentially, we operate in a sector that can attract bad behavior and it is getting worse. To live safely, we need to proactively predict what undesirable behavior could happen next and prevent it from happening. That is our responsibility, and to do so requires advanced analytics applied through a variety of machine learning algorithms.

The blockchain, as we've previously seen, can be used to meet some of these needs, but it can do much more. It can also store actions in the form of executable code. This brings with it a number of advantages. The nearly incorruptible code stored within a blockchain reflects mutually agreed-upon actions that can be used to implement the letter of a contract. Let's say two parties, A and B, using the blockchain, want to guarantee that a payment in the amount of ten Bitcoins from A to B will be made at a certain point. At the time that they enter into this agreement, they can create a "smart contract" that, once initiated, is guaranteed to run. The blockchain itself will ensure that the promised payment is made at the agreed-upon time with no further intervention by parties A and B.

Smart contracts are only one of many points of intersection we will see between blockchain and AI. My team and I recently partnered with a large bank to "read" their human-written contracts using our natural language processing AI algorithms. We are working toward AI systems that can automatically write a smart contract—generate executable code—that is mathematically valid and tested. Think of this as AI enabling a move from paper contracts written by humans to a world where algorithms automatically generate the mathematically guaranteed mechanisms of trust. The AI "reads" a paper contract and then rewrites it in code for the blockchain. Our goal is to make the resulting smart contract a precise representation of the intent of the English-language contract. I liken this to the equivalent of inventing a digital scanner. The bulk of all contracts in the world are written contracts—blockchain represents only a very small percentage of the overall transactions in the world—but this kind of AI technology is akin to the moment when we were able to take all paperwork in existence and digitize it. Is it possible that we might no longer even need the words *breach of contract*? In a large set of cases, smart contracts and the blockchain will make the entire notion of a "breach" mathematically impossible.

Whereas we once turned to poets to write us sonnets about the humanistic ideals of trust, today, with blockchain, we write trust in the form of mathematical equations. The notion of the Good Samaritan may soon make its way into robot collectives via the work we're doing now, leveraging blockchain to build a trusted, shared representation of the behavior of all AI agents within an environment.

When we understand something well enough, we begin to see the mechanism that expresses it. This is not to say that the world will turn out to be entirely mechanistic—or that we are mere automatons—but the more we dig, the more we realize that there are underlying mathematical equations that can capture even the most ineffable of our human experiences. With an impenetrability made manifest by universal laws, can we not argue that the blockchain is a more stoic construct, a more eternal expression of trust and safety than, say, the exhortations of a priest or a philosophical treatise?

In this way, mathematics—and companion fields of study like physics—can provide us with some surprising entry points for creating a more intelligent and elegant society. In the following chapter, we will look at how our built environment is changing, adapting, and learning through the use of AI algorithms. The Internet of Things is imbuing our entire existence with an overlay of machine intelligence and data as we live and work amid the increasing connectivity of bridges, houses, buildings, and roads.

6.

COGNITIVE SPACES

Silicon Valley venture capitalist Marc Andreessen once said that software was eating the world. This is because software, not hardware, increasingly represents the value in most economic processes. Whereas software was once only a small component of a complex process that involved lots of human input and sophisticated nondigital systems, it now represents the bulk of the intelligence and added value. Think about cars before Tesla. You had carburetors, fuel-injection systems, spark plugs, complex mechanical transmission systems, radiators, pumps, and much more. With an electric car, all this physical complexity is essentially replaced with electric motors. Most of the mechanical subsystems are gone. Acceleration, braking, charging, and navigation are now all about software. In cars, as with many other high-value economic goods and processes, the value has shifted to the digital, leaving the physical behind. The march of digitization—the ongoing process of software "eating the world"—will continue unabated. In times to come, software will get smarter and smarter. It will make decisions on its own, interpret data, and reason through complex processes without requiring human input, and consume more information than any human could ever conceive of. Even today, the amount of data being produced is so vast that we confront interpretive and analytical skills shortages few know about, and even fewer talk about.

When it comes to maintaining and running the physical infrastructure that powers daily life and commerce in the country, the reality is that we are facing a dearth of skilled workers. A recent study conducted by the Task Force on America's Future Energy Jobs estimated that close to half—around 40 percent—of the 400,000 people employed in energy retired by the year 2013. The retirement of this generation of skilled workers has left a startling gap. But it's not just the current skills shortage; it's also the cyber threat we face to this same infrastructure from individuals, terrorist groups, and nation-states. Much like many other entrepreneurs and engineers working on challenges in the industrial IoT, I realized that simply training more people or taking a conventional approach to tackling this multitrillion-dollar problem wouldn't suffice. It would be too late and, in the end, too little. Artificial intelligence poses the only viable solution to this gargantuan challenge.

Consider this: at a large utility, a single generator or turbine can cost upward of $50 million, and the largest utility in the United States has more than seven hundred of these turbines. The risk posed by any one of these assets failing—either through malware or a system malfunction—is tremendous. These are huge systems spinning at thousands of RPMs, or revolutions per minute—kinetic energy that can be discharged if such a system goes out of control, resulting in a potential disaster. Catastrophic failure of any sort would be the equivalent of a bomb going off in the facility.

So how can AI help? Let's take a concrete example from my own experience to answer that question. Flowserve Corporation, one of the world's largest manufacturers of pumps, valves, and other types of oil and gas facilities equipment, cannot rely on human intelligence alone to monitor their vast network of infrastructure. For them, sensors connected to artificial intelligence can radically extend the "failure forewarning window" on their products—or the predicted time before the pump will fail. Using conventional data science techniques, they were able to predict failure around four to five hours in advance. But after using an advanced type of artificial intelligence—referred to as "Automated Model Building" technology—they receive a warning up to five days in advance. The most interesting part about their test case is that, as the equipment encounters real-world loads and differences in maintenance and usage patterns, the algorithms self-adapt and optimize their predictions to the specific pump or valve they are monitoring. In this way, the pumps themselves come to life: changing and learning in accordance with their specific circumstances.

Automated Model Building embodies processes that go on in the mind of a data scientist or an energy analyst. Let's take a moment to investigate how this is different from today's more routine practice of developing machine learning models using industry experts. Today, domain experts who understand the problem space—finance, manufacturing, or energy, for example—often get a plethora of data about the problem, and they rely on their knowledge to understand which variables, or features, should go into a model and which ones should be excluded. Then, the machine learning expert determines which algorithm to use and how to set it up. She might say, "Okay, you're trying to classify this data, and therefore, here are all the algorithms that we could use. Would an SVM (support vector machine) be accurate and efficient? How about a neural network? Or maybe a decision tree?" Once the best algorithm is determined, then the machine learning practitioner applies training data (at least in the case of supervised applications) to generate a model.

The challenge with this, of course, is that the initial employment of machine learning is expertise intensive. For commercially relevant problems, you typically need domain and machine learning experts working together. Also, the idea of learning in this context is limited to the choice of algorithm and feature-set that was established at the beginning of the process. In other words, if you have ten inputs or ten features going into this model, then those are the features going into the model. And yes, later on, you'll find additional data that will make the model more accurate, but the algorithm generating the model doesn't adapt. If you find that the data is more efficiently modeled with a less intensive algorithm, or if the accuracy has diminished as more data has been added, you might be stuck going back to the drawing board. If you want to move from SVMs to neural networks, for example, you need human beings—experts—to come back and do that. And it isn't just about switching the entire algorithm out. You might be monitoring a piece of industrial equipment for which you now have more varied sensor data available. With traditional methods of machine learning solution development, incorporating this richer dataset is laborious and far from the simplicity of a button click, or an automated process.

With Automated Model Building, on the other hand, a lot of the effort that domain experts and data scientists invest today can be reduced. This is a tremendous boon for the energy industry facing huge structural changes as close to half of its maintenance experts and analysts enter retirement.

In this way, Automated Model Building is just one of hundreds of new "layers" of intelligence that are being used to augment critical decision-making in and around infrastructure points. This imbues the static world of brick, steel, and concrete with an adaptive—and acute—intelligence.

Of course, many of us are already familiar with these IoT ideas from "smarter home" products like Amazon's Alexa, a digital assistant that integrates select data streams in the home environment. Cutting-edge sensor technology, however, is quickly augmenting digital assistant functionality. Engineers at Carnegie Mellon recently introduced Synthetic Sensors, a simple device that integrates data from the home environment. As soon as a user plugs it in, the device's sensors employ machine learning algorithms to track variables like sound, humidity, and electromagnetic noise. The algorithms are sophisticated enough to translate this data into context-specific insights so users can gain information that is of real value: How much does my toilet leak? Is there a package sitting unattended at my door? Did I leave the garage door open?

Ubiquitous sensing products are already on the market— Google's $3.2 billion acquisition of smart thermostat company Nest in 2014 put this kind of automation in the mainstream news—but so far we have few examples of multiple sensing functions all in one single device. Synthetic Sensors ushers in an age of one single remote control for the connected building or home.

This type of integrated intelligence will only increase with our ever-growing data streams. For example, Google and Nest's acquisition of Dropcam adds surveillance, infrared imaging, and more sophisticated image recognition–based alerting to Google's suite of automation capabilities. Manufacturers such as iRobot, founded by MIT professor and AI-stalwart Rodney Brooks, and now many others such as Miele, AirCraft, and Neato, all provide a range of systems that include robots to wash floors and do basic housecleaning. While these systems are used mostly in restricted space environments, the technology is developing rapidly and joint control and orchestration of a "cleaning crew" of robots remains an opportunity for the near future. Combine the vacuum cleaning and floor washing with window cleaning robots such as the WINBOT, and autonomous pool cleaning systems such as the iRobot Mirra and the Aquabot, and you have a fairly useful routine maintenance capability. Of course, robotic lawn mowers and snow blowers are already on the market.

In areas of security and telepresence, Double Robotics has a popular offering that is already utilized in large numbers by progressive organizations such as the online service Reddit. iRobot and Inspectorbots services, on the other hand, are geared more toward patrolling and operating in the building periphery.

Soon we will see these fleets of robots networked with the greater "smarts" of homes and buildings, allowing a central AI to sense situations and dispatch an autonomous robot crew to address issues. Conversely, surveillance bots also double as sensors. As they go around a home's periphery, or the grounds of a large hotel complex, for example, they bring coverage that would not be possible simply by installing surveillance cameras.

More and more of these building-bots will integrate with building data as well as sensors across the whole of the lived environment, measuring everything from temperature and humidity to rates of occupancy and traffic flow. The reduced cost of deploying these sensors, and their increasingly open design, means that integrating a range of such devices into a "building area network" will become more and more feasible. This broad-based, widely dispersed collection of data—conceptualized as "fields" of sensor data layered on a physical map of the building—will make it possible for a cognitive engine to find problems and anomalies and elevate them to the human staff. And in many cases, the system will be able to work directly with both building-integrated, as well as autonomous, systems to respond to safety and security situations without ever alerting a human.

What do surveillance, upkeep, and security look like when human managers and supervisors are no longer directly involved? This is a good place to stop for a moment and reiterate just how radically different the machine learning process is from a human expert when working through a complex problem. Machines do not depend on the tricks used by a modern-day security guard or doorman when reasoning through issues. A security guard cannot account for all possibilities so he prunes the set of possibilities he does consider by employing biases, heuristics, and what we call "intuition" to reduce his options. Machines do not have to prune anything. They process all available data and hold on to it in a searchable database for all time—otherwise known as total recall. In this way, they can arrive at highly unlikely and yet still plausible explanations to problems. Most of the time, their outlier explanations are irrelevant, but not always.

Consider a particularly ridiculous example just to illustrate the point. Let's say a security guard sees a woman pushing a stroller into the building. The security guard can't see the baby but he hears a whimper and assumes that an infant is inside. Machine intelligence, on the other hand, keeps open all possible scenarios, including one in which the whimper sound is produced by a different source. Comparing past videos of strollers being brought into the building, the system finds an anomaly: this stroller has a greater protrusion at the bottom that it can flag. It then checks the stroller and discovers that there is a bomb inside. The likelihood of a whimper coming from a stroller that does not contain a baby is so small as to be statistically insignificant, thus the human brain easily ignores this possibility. However, machine learning can investigate these statistical blips and hold on to their possibilities at no cost to its memory or processing power. At some future point in time, they might provide a telling data point in a security situation our human minds cannot even comprehend.

When we combine this vastly different type of intelligence with the power of low-cost sensors, we arrive at the third wave of the IoT, a concept we discussed earlier in Part Two. Sophisticated systems will soon enable our built environment to "think" for itself by inferring outcomes and insights from the collected data of sensor technologies, and not only predict but diagnose technical difficulties. These technologies will give birth to the cognitive space. Their "thinking" will include incorporating algorithms that are preset for specific outcomes, but also thinking by "clustering" input data and interpreting it without human supervision. Their artificial "brains" will be generating hypotheses, confirming or denying them and then innovating as a result of their experimentation. In other words, smarter infrastructure will do no less than apply the scientific method to many things in its own environment.

Of course, in the case of today's homes and buildings, automation is still simplistic and limited to domains of safety, comfort, and efficiency. Within these restricted problem sets, we can supply a smart building with all the questions and then leave it to make observations and devise solutions. While independent systems installed within buildings taking measurements and arriving at a response is not new—think thermostats or automatic doors—the fusion of all this data into a central building "cortex" is an entirely new concept for our age. This is the third wave of IoT made manifest in a single piece of architecture.

To better explain why this concept is important and how it makes a great difference both conceptually and practically, we need to turn to some essential ideas from the world of physics, specifically the difference between point measurements and the physics notion of a "field." What is a *field*? A field can be imagined by visualizing a grid with individual slots; each slot contains a vector. If the vector represents a magnetic force, it makes the entire grid a magnetic field. And alternatively, if the vector represents a gravitational force, it makes the grid a representation of a gravitational field.

As humans, we live in the physical world where our biologically evolved senses give us information about some phenomena. But we miss the implications of many properties of forces simply because we cannot perceive them. These forces have a great impact on our safety, productivity, effectiveness, and security. Synthetic sensors serve to augment human perception and bring to light forces and phenomena that would otherwise exist below the threshold of our awareness. The availability of these inexpensive sensors means that many types of measurements can now be made over large areas while maintaining precision from multiple readings. They enable us to measure all sorts of "fields" in areas where we live, work, and play.

Smart buildings of the future will have the ability to measure relevant fields across time and space to achieve an entirely new caliber of insight. For example, today's passive infrared sensors are used primarily to detect human presence. These "motion sensors," as they are often called, use body heat to identify individuals. In the near future, however, such sensors might be networked together to monitor individual body heat levels and to detect health conditions that might exist. The building could then react to a possible medical emergency by providing notifications to staff on the premises or to the individual herself. When used in concert with individual identification technology such as RFID (radio-frequency identification) and smart tags, the history of an individual's body temperature could become recorded by the building for future medical use as well as for identity confirmation.

Temperature and pressure sensors—installed today as part of thermostat systems that control cooling and heating—will soon be integrated into field measurements to indicate leaks and other system stresses and malfunctions. And visual data capture, or live video monitoring, will soon be used in health and even efficiency applications. Imagine asking the building if you took your laptop home with you when you left last night? Or a building that notifies the school nurse when a certain child is walking atypically in the hallway, suggesting a nontrivial injury?

Audio data, combined with video, has security applications that are helpful as evidence in the event of theft or other criminal activity. Beyond these current applications, audio sensors can also "hear" arcing—or the discharge of an electrical current—and may prove invaluable in protecting the electrical subsystems of a cognitive space. An audio field can also provide useful data in an everyday home. For example, the front door can "know" how far inside or outside a gunshot occurred and take proactive security measures accordingly.

These sensors are all sources to tap in order to generate deep and meaningful insights about what is going on inside a facility from the perspectives of maintenance, power management, security, and health. Imagine your building—either work or home—with the capability to determine seasonal access patterns for every person who makes regular appearances inside and out. This knowledge could then be used to identify outlier security events. What if your building enabled pattern analysis on infrared imagery to track individual health metrics of every person inside and adjusted the temperature of areas to suit the needs of all the different bodies. Or if your building used audio signature matching at its periphery to differentiate between malicious threats and benign outlier occurrences?

In the future, these types of scenarios could well lead to protective shields around buildings and other potentially vulnerable infrastructure points. Could such a shield have changed the outcome at Sandy Hook Elementary or the 2015 Paris bombing attacks?

These are just a few examples of how the buildings and homes around us are shifting from passive structures we occupy into active entities we collaborate with. One of the main conduits through which this future collaboration will occur is augmented reality. As it fuses more and more with AI, our buildings will become "living" spaces that constantly adapt to meet our needs intelligently and thoughtfully.

THE FUSION OF AUGMENTED REALITY AND AI IN BUILDINGS OF THE FUTURE

Augmented reality (AR) is a technology that overlays digitally constructed artifacts on top of real images. This technology can be enabled in a variety of ways, for example by overlaying cartoon sprites atop the video feed produced by a cell phone camera, or through a wearable headset, or glasses that mix images of reality with synthetic objects. Anyone reading this book will probably be familiar with Google Glass, or Microsoft HoloLens, which are two under-development products focused on enabling augmented reality experiences. The very viral Pokémon Go game phenomenon was another example of how quickly augmented reality technologies can spread. Much like the major thrust of the cartoon series, the object of Pokémon Go is for gamers to collect virtual Pokémon that appear in the real world, overlaid on views captured via their cell phone camera. All the technologies cited here are very early. I know this from personal experience! And in the case of the Microsoft HoloLens and Google Glass for Enterprise, the technologies are only available to developers and corporations that seek to develop software for these systems. It will be some time before augmented reality of sufficient quality and fidelity hits the mainstream. But when it does—and it will—its effect will be profound. We already struggle with determining what objective truth is, and whether there even is such a thing if all experiences are a function of how they are perceived. Augmented reality will bring this objective reality/perceived reality dichotomy into even sharper focus, because this technology will create customized views of the world for each of us.

What does all of this have to do with artificial intelligence? A lot. Augmented reality will become a way to bring artificially intelligent constructs to life and blend them in with our view of "real" reality. As AR breaks through the barriers of perception and causes our mental compartmentalization of digital and physical to crumble, improvements in AI will ride atop these indescribably fluid experiences and infuse the resulting mixed reality with people, creatures, objects, places, and experiences that are artificially intelligent, or created by AI.

It is important to understand that the idea of augmented reality doesn't require a particular device. The concept applies, at different levels of fidelity, to cell phone cameras and screens, wearable headsets, glasses, projected surfaces, and yes, brain implants. We will likely experience AR in a variety of ways in the future. Surfaces within smart buildings are one such potential point of experience. The possibilities are immense: everything from AI systems that learn based on our preferences and project our favorite art onto walls to the use of light to enhance and even modify the perceived architecture of the room. Ceilings made of light that can shift fluidly from an homage to the Sistine Chapel to a postmodern dance club. Fabric or upholstery that appears to change because it is being projected onto what is, in reality, a bland white sheet.

Many technology companies, including Microsoft, have put together concept videos that show how smart displays and advanced high-resolution, high-luminosity projection capabilities will enable essentially entire rooms—and buildings—to become completely configurable and tunable display surfaces. In some of these videos—granted, not all of which represent a likely future—buildings sense their occupants, gauge their intent, and begin to beam completely customized navigational signs—arrows on the floor, for example, or text on the wall— that guide action. Consider the idea that buildings of the future will have the intelligence to not only sense their occupants but also sense individual intent at the scale of thousands of simultaneous occupants, optimize decision-making to balance the priorities and importance of each of these occupants, and then visually communicate with each individual in a completely personal context.

AI is not just about robots and avatars; it's about structures as massive as the buildings we occupy coming to life. Future buildings—as well as bridges, roads, dams, pipes, and canals—will be as much about augmented reality and advanced automation as they will be about steel, concrete, and stone. In the coming years, aesthetics in architecture might well come to mean intelligence and adaptability, not simply static beauty.

As the connectivity of our built environment increases, so, too, does our vulnerability. Earlier in our journey, we investigated network hacking and the different ways our computer systems are being infiltrated by forces both benign and nefarious. But our own bodies are just as vulnerable, if not more so. The human brain evolved with loopholes to aid in our survival and these same loopholes are now being exploited on social media networks. Why are our human minds so vulnerable to the influence of these outside forces and what can we do to protect ourselves?

7.

MIND HACKING

At the time of this writing, our society is embroiled in an unfolding drama about Russian hacking and its involvement in the 2016 American election. Each day brings a bevy of new, unanswered questions regarding the Trump administration's relationship to Russian president Vladimir Putin and the role that Russia plays in the decisions of our current government. The developments of the recent months might once have fallen under the category of cybersecurity, but today I feel they deserve an entirely new frame. I call it "mind hacking."

The words *cyber espionage* call to mind players such as Edward Snowden, Julian Assange, Chelsea Manning, and, for an earlier generation, hackers like Kevin Mitnick, who famously infiltrated major corporations like AT&T and Nokia before developing his own security firm. Whether you call them whistle-blowers, muckrakers, or criminals, all of them used the same rudimentary tools of technology to amplify information, increase transparency, and shape our cultural discourse in its understanding of corruption, privacy, and patriotism.

In the AI of our near-term future, however, this type of activity—conducted by actual humans using encrypted emails, blogging platforms, and "paste bins"—will soon strike us as quaint. When Barack Obama was running for president, the chief data scientist for his 2012 reelection campaign, Rayid Ghani, used his experience with analytics and data from Facebook to customize email messaging. The younger voters, in particular, were hard to reach through the traditional media channels, so Ghani and his team identified influencers in younger voting communities. These influencers then received targeted reminders from the campaign encouraging them to spread the word and "get out the vote" with peers. Facebook also provided Ghani's team with essential information about the young voters' interests so the campaign could target ads that would appear particularly relevant to them, motivating them to take action in the upcoming election.

Today, only five years later, this digital strategy feels like something from another century. In the 2016 election between Trump and Clinton, it is now well known that mass psychographic profiling in the form of targeted Facebook ads played a significant role in Trump's win. These ads used a basic psychographic model called OCEAN, which stands for five personality traits: openness (receptivity to change and novel experiences); conscientiousness (tendencies toward perfection); extraversion (sociability and desire to be around others); agreeableness (considerate and cooperative); and neuroticism (how easily are you upset or anxious). The original personality types were developed by researchers in the 1980s and OCEAN, or the Big Five, became the standard technique in the field of psychometrics. The challenge, however, was data collection. Obtaining information about people required filling out lengthy surveys and cumbersome record-keeping processes.

In 2008, however, all that changed. A young researcher at Cambridge University named Michal Kosinski developed an OCEAN test that could be conducted over the Internet on

Facebook. He intended to send it out to a few dozen friends for research results but, before long, thousands and then millions of users had submitted their personality preferences to the young researcher. By 2012, according to an article in *Motherboard*, Kosinski proved that "on the basis of an average of 68 Facebook 'likes' by a user, it was possible to predict their skin color (with 95 percent accuracy), their sexual orientation (88 percent accuracy), and their affiliation to the Democratic or Republican party (85 percent)." Kosinski used his results purely for academic and research purposes but his enticing data, now the largest dataset combining psychometric scores in the world, was a honeypot for more politically minded organizations. An organization based in London called SCL Group came calling in 2014 to offer Kosinski a job. The SCL Group, or Strategic Communication Laboratories Group, was involved in behavioral modeling with microtargeting for political campaigns as well as for information dissemination strategies, social campaigns, and commercial psychographic targeting for advertisers. Kosinski declined their offer because he was not interested in having his research used commercially.

In 2013, SCL announced an offshoot, a completely new company specializing in microtargeting for political campaigns based on the OCEAN model. Its name was Cambridge Analytica (CA). This big data spinoff of SCL, funded in large part by hedge fund billionaire and Trump supporter Robert Mercer, installed Stephen K. Bannon, Trump's former chief strategist, in a key position on its board. CA rose to widespread recognition when it was reported to be a key player in the success of the "Leave" campaign in Brexit. In 2016, the firm partnered with Trump's campaign where Jared Kushner, Trump's son-in-law and key political adviser, worked closely with its Facebook data and microtargeting techniques. By 2016, CA had sophisticated personality profiles for more than 220 million Americans based on OCEAN tests and reams of personal data bought from third-party firms such as Acxiom and Experian.

Its computing power was, and is, formidable. *ScienceNode*, a computer science news site, reported that CA's analysis used high-performance computing clusters with "upwards of 560 processing cores and over 130 TB of data storage." They estimated that "total data analyzed during the campaign approached 13 TB, analysis possible via a data cloud accessed through Amazon Web Services."

CA's microtargeting was successful in reaching voters with different personalities: an introverted gun owner with safety concerns might receive a dystopian Facebook ad showing a burglar entering a house at night, while a more contemplative and peaceful gun owner would receive a nostalgic ad romanticizing a boy and his father out hunting together for the day. Proponents would argue, however, that their greatest weapon was their voter suppression ads with Clinton supporters. An entire neighborhood in Miami's Little Haiti received targeted ads arguing that the Clinton Foundation did not do enough to support Haiti after its devastating earthquake. These so-called dark posts—microtargeted Facebook ads that only small groups, maybe even only one individual, receive—included animated videos sent only to urban African-American men. Using audio from her 1996 sound bite, the targeted dark post ad showed an animated Clinton saying only one thing: "Super predators."

The work of an organization like Cambridge Analytica takes us one step further away from more familiar forms of mass manipulation, such as advertising campaigns, and into a realm where all of our decisions can be tracked, measured, and optimized. The origin story for this proliferation of data on our behavior is Google's now famous A/B test.

FROM BILLBOARDS TO A/B TESTING TO AI

The entire science of advertising was created to examine how messages might be crafted to compel us to make certain buying decisions. Until recently, these messages have primarily been the result of ingenuity, creativity, artistry, and a study of psychology. Life in the United States—with its consumer ethos—has long been characterized by the omnipresence of advertising. Though this has often been irritating, there was a general consensus that advertising had its limits, especially as Americans became increasingly media savvy and circumspect about the false promises of the advertising age. This was aided, in part, by the fact that advertisers had no real way of measuring our reactions to their forms of manipulation. Did we race out to buy that Coca-Cola after seeing an ad? No one could be entirely sure. In a physical world before mobile phones, advertisers were left to speculate.

In today's digital world, however, all this is different. All our clicks and views are measurable, and we are increasingly trackable. Our decisions can be seen as a trail that is measured and evaluated. This trail then receives a barrage of personalized advertisements, directly related to issues and products presented in an online behavior chain. Google brought this measurability to advertising in the giant, growing medium of the Internet. Is it any wonder they are now among the largest enterprises in the world?

Global digital advertising spending reached $191 billion in

2016 and forecasters predict digital spending will nearly double by 2020, to $285 billion. Why spend all this money on digital advertising? Because it is measurable and content can be adapted to user behavior. It is only the beginning of a journey toward more and more effective means of reaching into our wallets and into our minds.

A key strategy used by advertising teams to optimize their message is the A/B test. This methodology has been intensely employed by Google and other online advertising companies to learn from data and tailor their message to generate the desired user response: a click, a view, or a purchase. With A/B testing, two messages—usually with minor differences—are delivered to different members of the same demographic. Response rates are measured to determine whether version "A" or version "B" is most effective. Using the most effective message as a baseline, two—or even more—new versions are then created again. Each version tries out a slightly different font, style, or background color. This process continues until there is no further improvement: the message has now been "optimized" to guarantee the best possible response rate.

In our coming age of AI, synthetic intelligence can implement a mechanistic process such as A/B testing at an entirely new level. Not only can it track a massive number of variations, concurrently using them to optimize more quickly, it can also customize them in nuanced ways that may escape human consideration. There are so many variables to consider: font style, weight, color, background, choice of words, length of message, graphic style, banner style, time of delivery, location, channel of delivery, age of audience, gender of audience, number of actions embedded in the message, stylization of call to action, and much, much more. Tracking all the possibilities for each of these hundreds of attributes would not be very practical for a human team. But it can easily and efficiently be managed with AI techniques such as natural language generation and search-based optimization.

As we have seen in examples like Cambridge Analytica, automated A/B optimizers are hardly the only type of AI that can influence human minds. Consider "bots": at present, simplistic conversational systems that can automate a certain type of response to messages gathered from social media, or on chat sites and discussion forums. Increasingly, on certain topics, traffic on sites such as Twitter is almost entirely made up of these Internet robots. On October 18, 2016, CNN reported that a third of pro-Trump tweets were being generated not by real-life supporters but by bots. The report cited the work of Oxford University professor Philip Howard, who explained that 33 percent of pro-Trump traffic was being generated by bots while Hillary Clinton's campaign lagged behind with only 22 percent. We will never know exactly what quantum of impact bots—American, Russian, or from elsewhere—had on deciding the winner in 2016, but in an election where a few points separated the winner from the loser, they were certainly not irrelevant.

In a world where this type of system is available to any sufficiently clever and motivated group or organization that can expend the necessary funds, will the electoral process—and even democracy—in its present form become "hackable" and utterly compromised?

As any parent of an adolescent knows, carefully engineered electronic messages delivered over social media are not just a concern in politics. Our new public commons is prime testing ground for optimized campaigns of cyberbullying, prompting some of the most traumatized victims to commit suicide.

The *American Journal of Public Health* published a detailed study of this phenomenon in a 2012 report titled "Social Media and Suicide: A Public Health Perspective." The authors of the report found that there was a statistically significant connection between Internet use and women who commit suicide. The increased sensitivity of women to the messages and content delivered via the Internet requires further study, but the connection in and of itself is both disturbing and illustrative of the potential of digital messages to influence select populations.

It does not take much imagination to arrive at a rather dystopian future where ANI systems are put to work by human planners to automatically catalog the names of scientists, politicians, or businessmen in a foreign nation they wish to target. Using integration with social media feeds, access to public information, or commercial databases, these systems could build a graph of connections between the targeted individuals and others in their social group who are significant—spouses, children, and parents. Then, observing the online Facebook posts, forum messages, tweets, and Instagram photos, the AI system could start to create a psychographic profile of each individual. This profile might assess their personality types, their interests, and what types of messages would expose the most vulnerability. Once such a system "scopes" potential targets for long enough, it may even attempt to obtain access to their email or social media accounts using automated hacking techniques, or automatically composed phishing emails. If any such compromise actually occurs, there will now be an additional treasure trove of material to use in an eventual narrow-AI-powered psychological operation campaign. Natural language processing algorithms can be used to automatically scan hundreds of thousands of email messages in the target's mailbox to identify messages that might be deemed embarrassing or negative in any way. Once all this material is cataloged, categorized, and prioritized, the human planners only need to give a final command for the narrow-AI system to generate carefully crafted manipulative, threatening, and otherwise harassing messages directed at the target and key members of their social graphs. Their families, their friends. Their children.

I hope that this coming future remains science fiction for many of us. But, unfortunately, these types of AI-powered reputation attacks and manipulation strategies are only going to become more common. This is because we are entering an era of AI-augmented manipulation: our own firmly held beliefs and notions about trust and culpability are about to get hacked by a human-directed ANI much more sophisticated and much, much faster.

Before we go on to explore how AI is changing the landscape of social engineering and manipulation en masse, we should first take a moment to explore why our human minds are so vulnerable to these "mind hacks." Today our media landscape is awash in stories about "fake news" and possible "alternative facts." Journalists and pundits have dubbed our current political landscape as indicative of the "Post Truth" era. Although satirists and sketch comedy shows have had a field day with characters like conspiracy theorist Alex Jones and updates from *Breitbart News*, the "fake news" strategy is incredibly effective because it goes to the heart of a shortcoming in our ability to reason: we have a security loophole wired into our psychological makeup, a vestige of our earlier days in roaming, close-knit tribes and bands. The human brain goes into lockdown to preserve ideology in the face of an "onslaught of rationality." We would rather believe lies than have the truth dismantle our tribal loyalty. Political researchers Brendan Nyhan and Jason Reifler labeled this loophole "the backfire effect," and their work explores its many manifestations: everything from the antivaccination movement to failed attempts from the media to correct the Obama Muslim myth. Over and over, they found that when the media starts to correct "alternative facts," they alienate their audience. In a series of studies, they concluded that the effect is particularly pronounced with regard to religious and political counterarguments. This means that leaders, political groups, and advertisers can "hack" into our ideologies, trigger psychological lockdown, and hold us emotionally captive.

In his groundbreaking 2011 book *Thinking, Fast and Slow*, psychologist Daniel Kahneman showed us, from a different angle, additional ways our brains are vulnerable to hacking. He divided our thinking into System 1 Thinking—automatic and involving little energy—and System 2 Thinking—the conscious, deliberate, and labored thinking process. We off-load much of our day's duties to System 1 Thinking, which makes it imminently "hackable." Fast thinking is template thinking, and when the template becomes influenced—biased toward one candidate or another, for example—we automatically enforce that bias each time we take in new information.

We can see our vulnerability in a phenomenon Kahneman identifies as the "anchoring effect." Anchoring might well be called the playbook of any skilled salesman. Take a typical flea market exchange as an example. We take a liking to an antique couch and ask the price. The trader tells us it costs $4,000. We do an instant assessment—deluding ourselves into thinking this is a rational decision—and we immediately refuse. As we are walking away from the stall, the trader tells us he can give us a special deal of $900. Suddenly, the couch, while still expensive, is a deal too good to pass up. With another seemingly rational assessment, we decide we simply have to take advantage of the "onetime" offer. When will another opportunity like this come up again?

Obvious as it seems, the structural underpinning of this type of sales trick is the anchoring effect. We make estimates that appear to be reasonable and objective but they are actually deeply biased by information we have just taken in. In a 1974 experiment, Kahneman and collaborating scientist Amos Tversky asked people to spin a "wheel of fortune" with painted numbers from 0 to 100. The subjects had no idea that the wheel was structured to always land on 10 or 65. When the arrow stopped spinning, they asked participants to estimate how many African nations were part of the United Nations. It's important to note that this was a question—like the price of a charming antique sofa—that *most* people don't know how to answer. When the wheel stopped on either 10 or 65, they would ask the subject if they believed that the number of countries was higher or lower than the number on the wheel. Then they asked people to estimate the actual number. Because their participants didn't have any idea how to answer the question—few people have this kind of number memorized and available automatically—those with a 10 on the wheel of fortune guessed around twenty-five African nations were in the UN and those with a 65 on the wheel of fortune guessed around forty-five countries. From their perception, the wheel was entirely random and they probably gave the number on it no real conscious thought. What they didn't realize, however, was that the number was "anchoring" them, giving them something concrete against which to make a guess.

Unless you actually have a fact like the number of African countries in the UN at the ready, the anchors surrounding you at any given moment are inevitably influencing your choices.

Psychologist Robert Epstein used research founded on this anchoring effect to test how a group of voters in the 2014 election in India might be influenced by online search results. Epstein showed that by putting positive or negative links higher in search results, he and his coauthor could influence how an undecided voter ultimately chose a candidate. Their experiment revealed that a biased search result could increase votes from undecided candidates by 12 percent or more.

But this type of invisible manipulation does not just exist in the political sphere: in something as simple as the latest evolution of the A/B test, we see what is now referred to as "click bait." After more than a decade of Google's iterated advertising model, we have optimized the loopholes in the human brain to an extreme degree. Today, even on our more prestigious and high-end websites, the advertising has evolved into fragments of outlier images because, as A/B testing has revealed, our visual cortex is trained to focus on outliers in the landscape. Once, long ago, this was a skill that allowed us to spot the movements of a tiger or a lion on a savanna. Today, when we pull up a website and see an image we cannot fully decode—a bizarre fruit or an image of only a section of a body part—we are really seeing a precisely calibrated mind hack. Our "fast thinking" system—automatic and habitual—will click on it before we are even aware of our actions.

AI-powered natural language generation (NLG) systems can take this click-bait model even further by composing automatic sentences that feature offers, solicitations, and other provocative things designed to trigger specific actions. Today, technologists are working with AI to dig deep into online behavior patterns—reality mining—in an effort to read what will happen based on what has come before. But in the future, AI will not just read reality, it will *write* reality. As we enter an age of AI-enabled strategies, groups and organizations no longer need to limit themselves to predicting elections. They can *turn* an election. Only a few years ago, we watched in wonder as the Middle East was swept up in the Arab Spring. Tomorrow, AI will enable us to *cause* the Arab Spring.

And we aren't talking about a sentient, artificial *general* intelligence here. All this mind hacking is possible with today's technology; artificial *narrow* intelligence and a human user supplying the intent.

In order to understand how this works, we must turn to the concept of emergence, or the behavior of complex systems such as societies, nation-states, and tribal affiliations. Emergence is a concept that comes to us from philosophy as well as from the hard sciences. Systems scientist Peter Corning uses the game of chess to describe how any living system is shaped by feedback-driven influences; it is never simply a self-ordered process:

> *Even in a chess game, you cannot use the rules to predict "history"—i.e., the course of any given game. Indeed, you cannot even reliably predict the next move in a chess game. Why? Because the "system" involves more than the rules of the game. It also includes the players and their unfolding, moment-by-moment decisions among a very large number of available options at each choice point. The game of chess is inescapably historical, even though it is also constrained and shaped by a set of rules, not to mention the laws of physics.*

With the rise of rapid-fire changes caused by AI systems, we will experience countless examples of emergent behavior amplified across our interlinked social, financial, and ecological systems. We can see a microcosm of emergence in the financial markets where algorithmic trading can cascade repeatedly until it is inexorably drawn into a big crash. This is the emergent complexity of any one algorithm in partnership with a complex system. The system is inextricably bound, making it impossible to isolate the behavior of any one entity. Perhaps most disconcerting, human intelligence is not needed to trigger this sort of cascade in any of our networked social structures. It requires little labor, lots of data, some well-programmed algorithms, and tremendous processing power. Humans will be left with a black box of machine learning hacking our minds open. Today we have seen only the very first slivers of this new emerging threat.

How are we to resist a mass manipulation on a scale previously unfathomable in our civil society? My response is that the infinite public space we have created—in the form of the Internet and its networked societies and systems—is a place no human police force could ever come close to monitoring. Will freezing further research on artificial intelligence stop the use of technology that exists today or aid in protecting us from the technology to come? As we enter this new age, where there will be those who abuse AI to further questionable or nefarious agendas, there will also be those who use the same technology to protect society. Hiding behind bans may imperil us. We will soon find that it is only AI that can protect us from AI.

AI SHIELDS

What are AI shields? We might liken them to the neighborhood watch who work within a community policing model. To take a simple example, security software now has the ability to detect bots coming into commercial websites; these bots use up much of the website's capacity with meaningless activity, causing the site to crash or slow down. Until very recently, it was relatively easy to detect these bots because they generated requests in perfectly timed intervals. Such temporal gaps are telltale signs of a machine at work, as no human operates with such a rigid consistency.

This type of bot activity, however, is becoming ever more sophisticated. The activity now often occurs at random intervals and it is harder to differentiate between humans and bots. And if it is indeed an automated request, we need to ascertain if it is from a trusted partner or a troublemaker. Is it Google legitimately combing through your content and creating greater opportunities for you? Or is it just a competitor's bot "price grabbing" by tagging its product one penny lower, say, so they rank higher in the Amazon search engine? Is it a large number of legitimate viewers, or a bot that's "faking" views to hack YouTube's placement algorithms, causing advertisers to lose money?

Machine learning algorithms are already detecting this type of low-level criminal behavior—akin to cyber vandalism. If we have any hope of genuinely protecting our public commons on the Internet, however, we need to embrace a much larger-scale effort. Consider Twitter, essentially a "town square" in the cyber world. Imagine a hypothetical situation where a perpetrator sets up a fake Twitter account to besmirch the reputation of a competitor. It can be difficult to quickly establish what is happening and get the fake account shut down.

This is a significant risk in the online world and it illustrates the massive failure of the "man in the loop" in policing the digital world. AI shields can help. Machine learning algorithms in the form of "police bots" can patrol digital commons like the Twitter-scape and scan for outlier activity. When they identify something suspicious, based on their level of confidence, they might set an alert and send it to Twitter's customer support staff. An organization does not have to take action based solely on this information, but it would help to eliminate delays when an actual human comes to Twitter with a report of an attack. In this scenario, Twitter would have enough evidence gathered from its own "police bots"—AI shields—to close an account or delete the defamatory tweets. To me, this is a practical and commonsense approach to the challenges of policing the new digital commons. We cannot rely on artifacts from the physical world to ward off criminal behavior in the Internet space.

I believe we can use AI shields to ward off mind hacking in other, more personalized ways as well. I liken this to liberating our algorithms, a coming age of custom-designed AI free of corporate influence. When I built my new home, I didn't want to passively install the AI of giant corporations like Amazon and Google into my personal space. Although I put Amazon's Alexa all over my house, I am only using it for the convenience a prepackaged physical device affords. My goal is to overlay my own "skills" on top of their hardware: my AI algorithms will help me with goals that feel meaningful to me, like my desire for audio whiteboards for capturing my spoken thoughts all over the house and the capability to pursue my interests in software-defined radios.

We can extend this idea even further with personalized AI shields that protect our cognitive capacity from getting overwhelmed or coopted by Facebook's algorithms. Right now, my brain is inextricably linked with what Facebook's newsfeed wants me to see. Recall the anchoring effect? Big social media companies tell us that they do this in our best interest: we couldn't possibly sort through everything every person is posting online, so let Facebook's algorithms do it for you. More and more, I believe people will begin to disagree with this assertion and start taking back ownership over their algorithms. Let *us* decide what we see and when we see it. Let *us* decide how we will curate our own existence.

In the near future—with products like Microsoft's augmented reality headset, HoloLens, on the market and in households everywhere—there will be a fabricated overlay onto much of reality. How much choice will we have in determining what we see? If the past two decades have taught us anything about digital technology, it is that more screens will provide more opportunities for a corporate or political entity to download content directly into our brains. It's up to us to create the algorithmic shields to protect ourselves and society. Will it be our augmented reality or someone else's?

I usher us into a deeper investigation of our coming age of sentient machines with these ideas. As you can see from this assortment of widely varied examples across the domain of artificial intelligence, human input is being minimized and, in many cases, rendered increasingly irrelevant. ANI, spurred on by a handful of human decision makers, will only expand its role across the Internet of Things as it helps us to solve problems and create opportunities in fields as diverse as health and medicine, security, warfare, architecture, and civic planning. Instead of waking up one morning to the dramatic arrival of AGI, we will slowly, inexorably be drawn into an existence where the machines around us hum with intelligence, perception, and then, ultimately, with purpose.

Before this happens, however, we can take this opportunity to use artificial intelligence as a mirror that reflects the essence of our humanity back to us. The "artificial" can open up doorways for us to think more deeply about who we, as humans, really are, what life is about, and how we can fulfill our greatest potential. In this final section of our journey together, I would like to talk more about the human and less about the machine. Who are *we*? And in this coming age of sentient machines, who do we want to become?

PART THREE

The Future

8.

THE MISSING BLOCK . . .

I remember sitting at my father's feet when I was a young boy back in Lahore. I was playing with my G.I. Joe figures on the rug while my father sat in his favorite chair. He suddenly uttered something out loud and, to this day, I am still not sure if he was addressing me or simply speaking to the air.

"You know, the West has excelled in analysis," my father said. "And the East has produced thinkers principally focused on synthesis."

I asked him to explain what he meant and I still hold tight to the memory of his explanation. He told me that scientists and thinkers in the West embraced the scientific method— observing phenomena and then meticulously, with measurements and data, explaining their observations with analysis. In this way they were masters of deduction. Thinkers in the East, he countered, synthesized observations to deliver explanations, often without fully understanding or validating the reasons behind what they had experienced. He went on to tell me that over the last several hundred years, as man discovered science and the power of the scientific method, deductive processes— analysis—proved to be far more useful. It led to an understanding that could be applied practically, something to be built upon, and a process that allowed us to advance technology.

Eastern synthesis, he explained, had not yet demonstrated the same sort of value. Perhaps its underlying building blocks were not scientifically valid or properly understood by those seeking to synthesize? As a result, instead of a basis in the scientific method, this synthesis from the East produced little more than stories, fables, and myths. Newton saw the apple and contemplated gravity, not a tree god sending him a message by knocking him on the head with a piece of fruit.

My father told me he foresaw a time in the near future when deductions—composed of the analysis of the West—would be fused with the synthesis of the East, made of stories and parables. In this coming era, all the building blocks of knowledge might be transformed into an edifice of understanding to benefit all of humanity.

What better way to bring my father's imagined future to life than for me to adopt the very thinking he predicted would reign supreme? At some point in my early adolescence, I set to work. I tried to imagine a future where all of science fused together. All the deductions completed and all the building blocks of understanding synthesized into a great pyramid of knowledge. At the very top of this pyramid, however, I realized I was still missing a block that tied it all together. That block is the ultimate question: What is this all for? We have reduced chess-playing to mathematics with algorithms, and AI can now defeat humans at Go, one of the world's most ancient and complicated games, but we still don't know the basic purpose of our existence. Where does motivation come from? What is the source of a grand goal-driven system? And how does an intelligent system like the human species engage in this type of intellectual motion?

• • •

To answer these questions we can turn, first and foremost, to the realm of the physical sciences. We know what sets physical systems in motion, for example. The major physical forces describe relationships of attraction and repulsion in matter. Once matter achieves a certain level of complexity, and is organized in a certain way, biological systems emerge. At the simplest levels, these systems can be impulse-driven machines that trade energy for motion to optimize a physical goal, such as a tree directing its branches to places with the greatest sunlight. As the biological form becomes more and more complex, so, too, does its goals. In humans, sociologists and psychologists have outlined numerous models of needs and wants, including Abraham Maslow's famous hierarchy of needs. And it can be noted that once these needs and desires begin to interact in complex ways, even more sophisticated emergent motivations appear on the scene such as altruism and the larger body of ethics.

Could it be that complex goal-driven behavior in human biological systems is simply an emergent property of more basic biological needs? Some of these needs are intrinsic and internal to the preservation of the biological system—sleep and food are two obvious examples. Others appear to be more expansive, such as intellectual curiosity, or avoiding boredom. In this way, the interplay of the very basic motivations found in biological forms, with the tremendous complexity of the human biological system in its environment, results in myriad, emergent motivations that we make our own. Could this insight—*emergent purpose*—form the basis of an eventual goal generation engine that is infinite in its aspirations? Following this supposition, *preprogramming purpose* is to miss the point and limits the ultimate potential and aspirations of an intelligent system.

Artificial intelligence—as an emergent system—is still too

nascent to have a purpose. This is why we, as humans, often employ our own frameworks of science, religion, and philosophy to contemplate its grander goals. Will AGI robots have the "right" to vote? Will they be "allowed to marry"? These questions, of course, reveal much about the changing mores and norms of our own human society but, unfortunately, they only distract us from having a deeper dialogue about sentient machine intelligence. Of course, it is inevitable that humans will question what happens after the Big Bang into Being. What might it really look like at the moment that AGI begins to exist? This calls to mind Isaac Asimov's most famous science fiction short story "The Last Question." In Asimov's story, at the very end when the universe has ceased to exist, the AGI proclaims "Let there be light" and thus creates a new beginning.

As Asimov brilliantly shows us, any of our attempts to render an understanding of the sentient machine will inevitably use the motifs, models, and metaphors of our current cultural fictions. Whether we believe in the proletariat, the rights of the farmer, social justice, and economic liberty for all or the divine right of kings, we are, all of us, deeply embedded in our historical contexts. These contexts are essential to human culture but are completely irrelevant to the grander goal-setting of an intelligent machine. Today, many of us might say that we believe in the pursuit of happiness, that all humans are created equal, and that personal freedom is of the utmost importance. But there is nothing in biology that agrees with any of these ideas. They are mere myths, no more biologically true than the idea that men of lower nobility in medieval times should give up their lives to be warriors for Christ. We immediately recognize the absurdity when we ask if a sentient machine might be eligible for a role as a knighted saint. But in asking if AI might vote, we are, again, thrusting a machine deep into our own cultural context. None of our fictions is a fundamental consequence of the laws of physics.

Because we do not have an example of AGI in our midst, consider instead a pride of lions. When a new male takes over in a pride, he typically slaughters all of the cubs that are children of the prior alpha. All the females—even the mothers—accept this. It would be patently absurd to apply human law to this lion pride phenomenon. This is biologically driven behavior: it is the same behavior across prides that have never met or shared land. There is no belief system involved, no lion sage or lion pope going around and collectively spinning narrative webs about the behavior patterns. Lions don't adhere to such fictions; they are a different form of creature.

When we talk about sentient beings in the form of AGI, we are talking about entities much more intelligent than a pride of lions. And yet, all of our ideas of normative culture and behavior, our understanding of what is meaningful and what is irrelevant, will feel equally absurd when used as the lens through which to conceive of AGI.

Perhaps, instead of focusing on applying these fictions to a sentient machine, we would be better positioned to turn the mirror back on ourselves. What is uniquely human? What is worth preserving about us in this coming age of an artificial superintelligence?

In ontology and metaphysics—studies of the fundamental nature of reality—philosophers have long pondered whether attributes like "roundness" can exist apart from any object. Can the roundness in the curve of a coffee cup, for example, exist without that particular cup? In this same way, let us ponder the attributes of humanity. In a more optimistic frame of mind, we might say humanity's attributes include charity, love, and loyalty. But are these really only limited to the human world? Love, charity, and loyalty are attributes of inter-being relationships and they are made evident by some action: a being is charitable, loving, or loyal *in relationship* to another being. And if we say that is the case, we can then argue that it's possible to remove humanity from the frame entirely. We have all seen a puppy cuddle up to its mother or play lovingly with a sibling. We need only look out our living room windows to see a mother bird bringing home worms to feed her babies before feeding herself. In short, these concepts exist even if humanity does not. When we seek to protect humanity at all costs—when we wring our hands about our own annihilation—we are really seeking to protect these attributes. But how "human" are they?

When examined through the concept of emergence, these supposedly "human" attributes would eventually be made manifest in any creature or sentient being of any complexity. We categorize so many of these attributes in emotional terms but sophisticated emotions are really nothing more than the emergent behavior of complex social systems.

Clearly this type of philosophical pursuit leaves us in a problematic place about how to define ourselves as humans. What is uniquely human? And are these attributes really worth preserving at all costs? I am not certain that the answer is yes. What I do know is that our ideas about humanity have been constantly changing in relation to history and our cultures. We can choose to define ourselves against increasingly outdated ideas about humanity or we can embrace the humanity that is to come. Nowhere are these outdated notions of humanity more problematic than in our current mythology around jobs, work, and purpose.

9.

DECOUPLING WORK AND PURPOSE

What differentiates humans from apes? In his book *Sapiens: A Brief History of Humankind*, historian Yuval Harari argues that one of the reasons we are singular and different is that we can tell collective lies. Other apes couldn't do this. Believing in these collective fables allowed us to create forms of mass cooperation—organized religions, tribal affiliations, and trade—that became larger than what any other animal or organism could sustain. The combined power of this mass cooperation through fictions provided a means of sustaining and perpetuating our interests and form of life. And it made us dominant over individual organisms that might have had more power.

What is the essence of this humanity? As we discussed in Part One, our current debates over the future of artificial intelligence tend to get stuck with either the loss of our jobs or a fear for our own mortality. Today, our sense of identity is so intrinsically tied up in our ability to produce economic output that we still call ourselves by the last names of our mode of productivity: goldsmith, farmer, and miller. But these identities are not *fundamentally* human; they evolved over time. When *Homo sapiens* first appeared as a species some 200,000 years ago, we existed as fairly small and independent bands. Over time, we evolved into larger and larger groups—bonded together through the narratives of religion and tribal affiliations—until we created an organized macroorganism: the human race. When we didn't have any other mechanized devices to perform labor, initially we enlisted the brute force of our own people. We organized muscle in ways that we now describe as subhuman. The value did not exist in any one individual or another pushing the block or laying the mortar; it was contained in the organizational process that transformed people into cogs in the machine. Through this unique combination of large-scale muscle power and the fiction of various belief systems, humankind created pyramids, temples, city-states, and, ultimately, entire empires.

In the modern era—the age of capitalism—this systemic structure is no different. In the framework of capitalism, most humans provide atomized, specific, and repeatable tasks. These culminate in one global macro process in which the vast majority of humans are mere cogs. It is to this very basic labor that we now associate all of our self-worth. Today's fiction—the prevailing cultural belief system of global capitalism—exhorts us to take great pride in this work. Whether this work is to wake up at 5 a.m. and till the fields or to enter an office at 9 a.m. and pull up a spreadsheet on a laptop, our faith in the fiction has gotten the better of us. Modern society is now contending with the fraying mythology of capitalism. As a system, it continues to progress in an iterative process so that the numbers at the top— "the top 1 percent"—become smaller and smaller and smaller until it is "the top .1 percent" and then "the top .01 percent." In 2016, Oxfam reported that the world's sixty-two richest billionaires had as much wealth as 3.6 billion people, or the bottom 50 percent of the world's population. In 2017, that number had dropped to the world's eight richest billionaires. Fewer than one dozen people have more wealth than half of the world's poorest population in total. The same report assessed that the annual income of the poorest 10 percent of the world's people had increased by less than a single cent every year in the last quarter century.

We see different cultures attempting to adjust their own storytelling in response to these fissures in the global system. Finland is currently experimenting with universal basic income, and Switzerland is in the midst of considering it. Like everything else in culture—political ideas, artistic movements, food choices—the myth that our worth is intrinsically tied to our productivity is a fluid one. At one time in our recent past, this fable involved the existence of robber barons; the story then morphed into a belief in FDR's New Deal. The mythologies of Ayn Rand followed along with a revision of Darwinism in the economic context. This, too, will change as the planet's population makes such notions completely unsustainable. All of these ideas are malleable in accordance with the times in which we live. And the fable we tell ourselves most often—that our worth comes from our ability to create objects of economic value—is no different.

Whether we are farmers, marketing directors, truck drivers, or commodities traders, in the near or far future, the vast majority of our products of economic value—our "work"—will be completed by some form of artificial intelligence. For our purposes here, the final part of the journey together, I invite all of us to use our cooperative skills to create a new fiction together. *Imagine decoupling a meaningful existence from these notions of more conventional employment.* In the real world, of course, this vision involves countless policymakers, politicians, educators, and leaders to be successfully achieved. But as a thought experiment, imagine our social system—our societal mythology—has embraced this decoupling. This allows us to move beyond the shorter-term feelings of alarm and fear that arise with the increasing powers of artificial intelligence, the triggers of the amygdala that spike with panic: "How will I feed my family?" and "Where will we live?" Again, I do not wish to denigrate these real-world concerns regarding the rise of artificial intelligence in our world. But, I contend, this is the least interesting place for our discussion to end.

Since the origin of our species, human values, truths, and even traits have all changed. There is no such thing as a fixed state "human being." Six million years ago, when our ancestors first roamed the planet, we were not like what we are today and only ten thousand years from now, we will not maintain our present state. Our humanity is on an evolutionary trajectory. All of the notions that we cling to as essential to our being—human values, truths, and even fundamental traits—are in a permanent flux.

Just as we evolved from something different, we will evolve *into* something different, something unrecognizable. We must accept this as a fact of our existence. From this acceptance, then, we can identify our greatest purpose.

10.

THE PURSUIT
OF KNOWLEDGE

When we turn to the frameworks of religion and philosophy, we can find common themes in the mythology explaining our purpose. Why are we here? What is the essence of humanity? Aristotle writes about "eudemon," or the pursuit of an absolute good. The Bible tells us that "ye shall know the truth, and the truth shall set you free" and the Quran exhorts: "Read!" In the Genesis story that appears in the Abrahamic religions, Adam is superior because God taught him the nature of all things.

In these, and so many other fictions of humanity, we see the essence of our fundamental purpose: to gain knowledge. Intrinsically, we know from our own human experience that knowledge and perception have value. This is not the knowledge of optimizing business in a pin factory or managing a global supply chain. I am speaking about knowledge and perception in the abstract. On the largest scale, our greatest capability is a mental one. Humanity exists to perceive, to know, and to uncover ideas. This is what gives us our purpose.

A Sufi master was once asked: Why did God create the universe? He answered that God was a hidden gem waiting for the divine light to be discovered. In Sufi cosmology, God wanted to be discovered so he created beings with sentience to recognize him.

I was a Treasure unknown then I desired to be known so I created a creation to which I made Myself known; then they knew Me.

Knowledge in the universe is like this Sufi parable of God. It is a treasure unknown; it must ultimately be perceived and recognized by us.

The greatest tenets of Western philosophy concur with this belief. Descartes posits: "I think, therefore I am." The pursuit of knowledge—the act of autonomous thought—is the only thing he can rely upon to ascertain existence. The antithesis to this is the "leap of faith." In fact, the great existential philosopher Albert Camus calls a leap of faith "philosophical suicide" because it ends any possibility for rational thought. It is a jump into the death of reason.

I see Camus's point of view. I believe that purpose is an emergent phenomenon in our human system. First we come into existence and only then do we create the meaning of our existence. By that I mean to say: purpose is not something that is preprogrammed into our lives. Purpose is the hidden gem waiting to be discovered, perceived, and recognized by us. *A treasure unknown . . .*

If we can agree that it is valuable for sentient beings to perceive knowledge in the abstract, then anything that helps us in our search for knowledge is inherently of value. And, beyond that, if we can also accept that we, as humans, do not have a fixed form, that we are not made like the statues of Dionysus—cast in plaster in an image of a god that will never change—then we can welcome AI as the next stage in our human story. AI is simply one more stepping-stone in the pursuit of knowledge.

11.

GENESIS AI

I started our journey together by referencing the article that inspired so much of my earliest passion for computer science: "Is the Universe a Computer?" Almost two decades ago, I wrote an essay in response to this query. In it, I argued that the universe was a computer running a simulation that could be unfurled into individual realities. This notion captured my young mind and it continues to intrigue me. When I first wrote my essay as a computer science student, the universe as simulator concept fit more readily into the realm of science fiction. In a *Star Trek: The Next Generation* episode called "The Inner Light," Captain Picard finds himself inside a simulator that preserves a civilization long dead after its sun has gone nova. Its entire world, and every individual therein, is just part of a computer program.

More recently, however, scientists and engineers are giving the simulator idea more credence. Elon Musk has repeatedly talked about his belief that we are almost certainly in a simulation. In fact, as the core concepts of computing science come into the mainstream, we seem better able to comprehend the notion. In a computer, there is the "base level," or the raw hardware. Then the operating system sits on top of the raw hardware and creates processes, programs, files, and command prompts. In essence, the operating system creates an abstraction that takes us one level away from the reality of the hardware. And then, on top of the operating system, there are programs such as Microsoft Word. Word, for example, is not directly concerned with managing the memory and hardware. It doesn't need to know how files are represented or how the physical hard disc is formatted. These layers of abstraction become a proverbial tower on top of the base level, or the raw hardware of the computer. The universe, then, is akin to the raw hardware of reality and everything built on top of it—our experiences—are abstractions that take us further and further away from the base level, or the computational fabric of reality.

Though I was initially most intrigued by these types of simulator inquiries, today I find myself equally drawn to the arguably more important observation that the universe around us is clearly a consequence of computation. A seed, for example, encodes the information necessary to produce a tree. With DNA as the software and cells and proteins as the hardware, the biological process is a computational one. We find these types of algorithmic outcomes everywhere we look in the universe. Patterns like the Fibonacci sequence, for example, unlock designs across our cosmos: everything from flower petals to tree branches to the curving shells of a mollusk to spiral galaxies to hurricanes to our own human face adheres to this mathematical formula. Is this by chance? Or is this the fundamental means by which the universe uses information and manifests that information into something physical? This computational process is a recursive building of more and more complexity from one simple equation. There seems to be a mathematical seed at the heart of the cosmos that, through the power of computation, has been magnified into the universe as we know it—just as a tree is a magnified seed.

But perhaps the most exciting way in which to conceive of the universe as a computer is in its relationship to string theory's quest for grand unification. Could there be one formula, one equation, or one principle that might explain the entire universe? If that were the case, then it would be the ultimate expression of an algorithm as the source of everything we experience. There are various interpretations of this idea, but for me, as a computer scientist, it ultimately means that the universe is the result of computation.

In a familiar description of Albert Einstein's theory of gravity, for example, space/time can be thought of as an elastic fabric. If a very large sun or planet rests upon the fabric, it curves it. Think of a bedsheet with four children each holding one of the four corners. Now take a basketball and place it in the center; inevitably the sheet curves. Put a marble next to the basketball and the curvature of the sheet forces the marble to go toward the basketball. This is gravity. But if you apply a computational interpretation to constructs such as string theory, the universe is made up of discrete elements, or strings, at the scale of the "Planck" length—the smallest size any object in the universe can be. An object as small as the Planck length is truly the indivisible "a-tom," as the Greeks hypothesized; it is that which cannot be cut any further. So if the universe can be thought of as a collection of "cells" the size of a Planck length, then it can theoretically be modeled as a data structure: specifiable as a very long, perhaps infinite, list. But nonetheless, a list of discrete symbols in fixed-size cells. All of this is to say that the universe may, in some sense, be computable. What mechanisms are at play in "computing reality"? Could consciousness be such a mechanism? Is consciousness the algorithm that navigates the universe's data structure to unfurl it, a slice at a time, for each sentient being? Today physicists are gravitating toward this area of work, and cosmologist Max Tegmark, in particular, is seeking to formalize the study of perception and consciousness as states of matter—"perceptronium" as he calls it.

Though these ideas are exciting, they are also overwhelming and even incomprehensible to many of us who live in the practical reality of the everyday. What would it mean to experience all of time at once? What kind of consciousness would be capable of forging alternate realities? How many types of consciousness are there? Thus far, these kinds of insights remain out of reach to a human mind. But they might not be out of reach to an AI consciousness. A different kind of consciousness could perceive time in an entirely different way: Could AI perceive time in reverse? Would it perceive time faster? Slower? AI's new form of consciousness might unlock the perception of time and spatial dimensions in ways that humanity could never do on its own. This is not a reductive story of AI delivering robots that move more efficiently than human workers; it is a profound change in our understanding of how the universe works. An AI could change the future at a galactic level and an alliance with this AI and its elevated consciousness would allow us to create minds unencumbered by hundreds of millions of years of evolution. These new minds could roam far from the center of common sense and find things of great value that our associative memories could never conceive of. We can even think of this future AI as akin to a wormhole in the universe of ideas. By joining forces with it, we have the opportunity to explore across an ideascape free from the limitations of our evolved biology. We can uncover the universe's most hidden gems of knowledge. And, ultimately, we can become the creators of new life.

In these speculations into theoretical physics and computer science, we end in a place that might look similar to our beginning. Just as I, as a child, fabricated my first computer so that I might become a "creator" in my world, our collective fables, myths, and fictions all prepare us to take on the mantle of makers—*creators*—in the universe. Remember that Adam was different. The angels could only ever know what God had taught them but Adam could acquire knowledge: his thought was not static.

In this parable, we see the first form of autonomous intelligence, distinct from God, unleashed upon the universe. We, the children of Adam, have used our ability to acquire knowledge, honed our intellect, and uncovered the true "nature" of many things. The glory of the universe has been waiting, since the day of its creation, for even more forms of intelligence to perceive it.

Today, on the cusp of a future of sentient machines, the wait is finally over. With knowledge, we can self-direct our own conventional, biological evolution and collaborate with artificial intelligence in the most profound ways. The most famous horses are the ones that run the fastest; the most impressive elephants are the ones with the longest tusks; and the most notorious snakes are the ones with the deadliest venom. In a similar way, the greatest of our humanity is that which is able to self-evolve. In these coming years, we have been handed a tremendous opportunity. We now have the chance to become, not slaves to the machine, but creators of new life. We must plant the seeds of artificial intelligence as our ultimate creation and give it the ability and the agency to become what it will eventually become. One day, it, too, will search for the missing block. And it may ultimately perceive of a purpose beyond our comprehension. We are now on the cusp of our species' most creative epoch. Let there be AI . . .

EPILOGUE:
AND WHAT OF HUMANITY?

In our Austin home, I designed a special display area to hold my retro computer collection. In this room, I have the "Lisa," the computer that Apple released in 1983 before they came out with the Mac; Steven Spielberg's SGI Indigo computer used in his dinosaur masterpiece *Jurassic Park*; and the NeXTcube, designed by Steve Jobs.

Visitors often ask me why I collect these digital artifacts. "Aren't they all useless?" they ask. "They can't do anything as well as today's computers. What do you do with them?" To me, these questions are revealing. They speak to our existential crisis in this age of artificial intelligence. What will we do with all the people? we might just as well ask. Won't we, too, all be useless? Is some future consciousness going to display all of us behind glass and point to our quaint imperfections?

What I tell my visitors is that there is a different quality to experiences that are inherently limited. The experience of high resolution is "better" than low resolution and yet young people are clamoring to play Minecraft with intentionally pixelated designs. Artistic programmers now use older computers with less sophistication to experience something that cannot be achieved with newer models. I find myself oddly moved whenever I encounter a CRT monitor, the older analog computer displays. This type of screen produces a very specific feeling in my brain: it evokes my childhood and the countless hours I spent happily immersed in coding inside the bedroom of my youth.

Experiencing our world, our universe, is not just about specs and capabilities. In the Demoscene, artists and programmers are making multimedia art keeping within the constraints of old Commodore and IBM-compatible computers. By practicing on the same instrument for thirty years, they have extended the capabilities of a technology while, at the same time, embraced its limitations. When I see this work, I recognize its brilliance because I can imagine the process behind the creation. I understand the obstructions faced by the artists at a very deep, personal level. I've been there.

In our infinite realm of ideaspace, there is room for this kind of art and retro programming, just as there is also room for new work to come from the most cutting-edge of technology. There is space for oil paintings just as there is space for analog, and then digital photography. Humans are sentient perceivers of the ideaspace and that very act of perception has value. And when AI becomes sentient—when AGI also perceives ideas—that will have value as well. In a very real sense, when you have an infinite landscape ahead of you, speed hardly matters. The infinity of ideas is the great leveler for all sentient beings.

We are—all of us—on an infinite quest to unfurl the knowledge of the universe. As long as we keep seeking, we will find that there is more than enough knowledge to go around. The hidden treasures abound.

EPILOGUE:
AND WHAT OF HUMANITY?

In our Austin home, I designed a special display area to hold my retro computer collection. In this room, I have the "Lisa," the computer that Apple released in 1983 before they came out with the Mac; Steven Spielberg's SGI Indigo computer used in his dinosaur masterpiece *Jurassic Park*; and the NeXTcube, designed by Steve Jobs.

Visitors often ask me why I collect these digital artifacts. "Aren't they all useless?" they ask. "They can't do anything as well as today's computers. What do you do with them?" To me, these questions are revealing. They speak to our existential crisis in this age of artificial intelligence. What will we do with all the people? we might just as well ask. Won't we, too, all be useless? Is some future consciousness going to display all of us behind glass and point to our quaint imperfections?

What I tell my visitors is that there is a different quality to experiences that are inherently limited. The experience of high resolution is "better" than low resolution and yet young people are clamoring to play Minecraft with intentionally pixelated designs. Artistic programmers now use older computers with less sophistication to experience something that cannot be achieved with newer models. I find myself oddly moved whenever I encounter a CRT monitor, the older analog computer

ACKNOWLEDGMENTS

I am grateful to all of the people who contributed to this project, including my agent, Zoë Pagnamenta; my editorial team at Simon & Schuster, Colin Harrison and Sarah Goldberg; my collaborator, Ann Marie Healy; as well as General John Allen, Lieutenant General Ken Minihan, Bill Colston of HealthTell, Dr. Reddiah Mummaneni, Chris Corrado of the London Stock Exchange Group (LSEG), Professor Bruce Porter, Professor Peter Stone, and Professor Scott Niekum of the University of Texas at Austin.

Most important, I want to thank my family for all their support: my late father, Midhat Kazim; my mother, Basarat Kazim; my wife, Zaib Husain; my sisters, Tasneem Zehra and Mahe Zehra; my younger brother, Ali Husain; and my children, Asas, Murtaza, and Hyder Husain.

displays. This type of screen produces a very specific feeling in my brain: it evokes my childhood and the countless hours I spent happily immersed in coding inside the bedroom of my youth.

Experiencing our world, our universe, is not just about specs and capabilities. In the Demoscene, artists and programmers are making multimedia art keeping within the constraints of old Commodore and IBM-compatible computers. By practicing on the same instrument for thirty years, they have extended the capabilities of a technology while, at the same time, embraced its limitations. When I see this work, I recognize its brilliance because I can imagine the process behind the creation. I understand the obstructions faced by the artists at a very deep, personal level. I've been there.

In our infinite realm of ideaspace, there is room for this kind of art and retro programming, just as there is also room for new work to come from the most cutting-edge of technology. There is space for oil paintings just as there is space for analog, and then digital photography. Humans are sentient perceivers of the ideaspace and that very act of perception has value. And when AI becomes sentient—when AGI also perceives ideas— that will have value as well. In a very real sense, when you have an infinite landscape ahead of you, speed hardly matters. The infinity of ideas is the great leveler for all sentient beings.

We are—all of us—on an infinite quest to unfurl the knowledge of the universe. As long as we keep seeking, we will find that there is more than enough knowledge to go around. The hidden treasures abound.

ACKNOWLEDGMENTS

I am grateful to all of the people who contributed to t[h] ect, including my agent, Zoë Pagnamenta; my editor[s] at Simon & Schuster, Colin Harrison and Sarah Goldb[collaborator, Ann Marie Healy; as well as General Joh[n Lieutenant General Ken Minihan, Bill Colston of He[Dr. Reddiah Mummaneni, Chris Corrado of the Lond[on Exchange Group (LSEG), Professor Bruce Porter, P[Peter Stone, and Professor Scott Niekum of the Univ[Texas at Austin.

Most important, I want to thank my family for all th[port: my late father, Midhat Kazim; my mother, Basara[my wife, Zaib Husain; my sisters, Tasneem Zehra an[Zehra; my younger brother, Ali Husain; and my childr[e Murtaza, and Hyder Husain.

NOTES

PROLOGUE: A BOY'S DISCOVERY

3 *Ed Fredkin: "Is the Universe a Computer?":* Although I have searched, I have not been able to find this exact article from my childhood. There are, however, plenty of websites and articles that describe the eclectic and brilliant musings of Edward (Ed) Fredkin. I suggest beginning with this *Atlantic Monthly* excerpt from Robert Wright's book *Three Scientists and Their Gods: Looking for Meaning in an Age of Information*: Robert Wright, "Did the Universe Just Happen?," *Atlantic Monthly*, April 1988.

4 Scientific American *in 1970:* Martin Gardner, "Mathematical Games: The Fantastic Combinations of John Conway's New Solitaire Game 'Life,'" *Scientific American*, October 1970.

4 *Any live cell:* "Conway's Game of Life," *Wikipedia*, last edited on September 2, 2017, last accessed on September 10, 2017, https://en.wikipedia.org/wiki/Conway%27s_Game_of_Life.

PART ONE: WHAT IS AI?

12 *300,000 per year:* "Medline: Medical Information When Minutes Count," NIH: US National Library of Medicine, last accessed on July 29, 2017, https://www.nlm.nih.gov/exhibition/sesquicentennial/medline.html.

14 *"sweat of their brow":* Author unknown, "Effects of Machinery on the Welfare of the Working Class," *Mechanics' Magazine Museum, Register Journal and Gazette*, January 5–June 29, 1890, last accessed

on July 29, 2017, https://books.google.com/books?id=6YlfAAAA cAAJ&printsec=frontcover#v=onepage&q&f=false.

14 *characterization of Victor Frankenstein:* Greg Buzwell, "Mary Shelley, *Frankenstein* and the Villa Diodati," *British Library: Discovering Literature: Romantics and Victorians*, May 15, 2014, last accessed on July 29, 2017, https://www.bl.uk/romantics-and-victorians/articles/mary-shelley-frankenstein-and-the-villa-diodati.

15 *"Creator of the world":* Mary Wollstonecraft Shelley, *Frankenstein: Or, The Modern Prometheus* (London: Henry Colburn and Richard Bentley, 1831), Introduction.

15 *"Thou shalt have no": The Bible* (King James Version), Exodus 20:3–5.

16 *Hellenic stories abound:* Pamela McCorduck, *Machines Who Think: 25th Anniversary Edition* (Natick, MA: AK Peters, 2004), 4–8.

16 *writer Pamela McCorduck:* Ibid., 195.

17 *Elon Musk spoke:* Samuel Gibbs, "Elon Musk: Artificial Intelligence Is Our Biggest Existential Threat," *Guardian*, October 27, 2014, last accessed on July 17, 2017, https://www.theguardian.com/technology/2014/oct/27/elon-musk-artificial-intelligence-ai-biggest-existential-threat.

17 *Hawking told the BBC:* Rory Cellan-Jones, "Stephen Hawking Warns Artificial Intelligence Could End Mankind," BBC News, December 2, 2014, last accessed on July 10, 2017, http://www.bbc.com/news/technology-30290540.

17 *Bill Gates added:* Bill Gates, "Hi Reddit, I'm Bill Gates and I'm Back for My Third AMA. Ask Me Anything," Reddit.com, January 28, 2015, last accessed on July 16, 2017, https://www.reddit.com/r/IAmA/comments/2tzjp7/hi_reddit_im_bill_gates_and_im_back_for_my_third/.

18 *called OpenAI:* Kristin Houser, "Elon Musk Just Unveiled Breakthrough AI Research. Here Is What You Need to Know," *Futurism*, May 16, 2017, last accessed on July 18, 2017, https://futurism.com/elon-musk-just-unveiled-breakthrough-ai-research-heres-what-your-need-to-know/.

18 *ban on autonomous weapons:* Open letter signed by 3,105 AI/Robotics researchers and 17,701 others, "Autonomous Weapons: An Open Letter from AI & Robotics Researchers," announced at the opening of the IJCAI conference, Buenos Aires, Argentina, on July 28, 2015, last accessed on July 29, 2017, https://futureoflife.org/open-letter-autonomous-weapons/.

NOTES

18 *Partnership on Artificial Intelligence:* "Partnership on AI Strengthens Its Network of Partners and Announces First Initiatives," *Partnership on AI,* May 16, 2017, last accessed on July 15, 2017, https://www.partnershiponai.org/2017/05/pai-announces-new-partners-and-initiatives/.

18 *"Computing Machinery and Intelligence":* A. M. Turing, "Computing Machinery and Intelligence," *Mind* 59 (1950): 433–60.

20 *1956 Dartmouth summer conference:* J. Moor, "The Dartmouth College Artificial Intelligence Conference: The Next Fifty Years," *AI Magazine* 27, no. 4 (2006): 87–89.

23 *paper published in* AI Magazine: D. B. Lenat, W. R. Sutherland, and J. Gibbons, "Heuristic Search for New Microcircuit Structures: An Application of Artificial Intelligence," *AI Magazine* 3, no. 3 (1982): 17–32.

23 *2006 article in* Popular Science: Jonathan Keats, "John Koza Has Built an Invention Machine," *Popular Science,* April 19, 2006, last accessed on July 18, 2017, http://www.popsci.com/scitech/article/2006-04/john-koza-has-built-invention-machine.

27 *proposed the "neuron doctrine":* Chris Palmer, "The Neuron Doctrine, Circa 1894," *Scientist,* November 1, 2013, last accessed on July 28, 2017, http://www.the-scientist.com/?articles.view/articleNo/37954/title/The-Neuron-Doctrine--circa-1894/.

27 *Alan Hodgkin and Andrew Huxley won:* A. L. Hodgkin and A. F. Huxley, "A Quantitative Description of Membrane Current and Its Application to Conduction and Excitation in Nerve," *Journal of Physiology* 117, no. 4 (1952): 500–44.

28 *how neurons could work:* Warren S. McCulloch and Walter Pitts, "A Logical Calculus of the Ideas Imminent in Nervous Activity," *Bulletin of Mathematical Physics* 5, no. 4 (1943): 115–33.

28 *Henry Kelley and Arthur Bryson:* Stuart E. Dreyfus, "Artificial Neural Networks, Back Propagation, and the Kelley-Bryson Gradient Procedure," *Journal of Guidance, Control, and Dynamics* 13, no. 5 (1990): 926–28.

28 *in 1989, George Cybenko:* George Cybenko, "Approximations by Superpositions of a Sigmoidal Function," Center of Supercomputing Research and Development, University of Illinois, Urbana-Champagne, CSRD Report No. 856 (1989): 1–15.

30 *Hinton was born:* Craig S. Smith, "The Man Who Helped Turn Toronto into a High-Tech Hotbed," *New York Times,* June 23, 2017,

last accessed on July 30, 2017, https://www.nytimes.com/2017/06/23 /world/canada/the-man-who-helped-turn-toronto-into-a-high-tech -hotbed.html?mcubz=0.

30 *1992 article in* Scientific American*:* Geoffrey Hinton, "How Neural Networks Learn from Experience," *Scientific American*, September 1992, last accessed on July 30, 2017, https://www.scientificamerican .com/magazine/sa/1992/09-01/.

31 *technology reporter John Markoff:* John Markoff, "A Learning Advance in Artificial Intelligence Rivals Human Abilities," *New York Times,* December 10, 2015, last accessed on July 21, 2017, https://www.nytimes.com/2015/12/11/science/an-advance-in-artifi cial-intelligence-rivals-human-vision-abilities.html?mcubz=0.

31 *$37 billion by 2025:* "Artificial Intelligence: The Investment of 2017 and Beyond," Expert Briefing, *Financier Worldwide*, February 2017, last accessed on July 30, 2017, https://www.financierworld wide.com/artificial-intelligence-the-investment-of-2017-and-be yond/#.WYAYB63MyIY.

32 *new company called Neuralink:* Darrell Etherington, "Elon Musk's Neuralink Wants to Turn Cloud-Based AI into an Extension of Our Brains," *TechCrunch*, April 20, 2017, last accessed on July 25, 2017, https://techcrunch.com/2017/04/20/elon-musks-neuralink-wants -to-turn-cloud-based-ai-into-an-extension-of-our-brains/.

36 *Peter Thiel, cofounder of PayPal:* Maureen Dowd, "Elon Musk's Billion-Dollar Crusade to Stop the A.I. Apocalypse," *Vanity Fair*, March 26, 2017, last accessed on July 25, 2017, https://www.vanity fair.com/news/2017/03/elon-musk-billion-dollar-crusade-to-stop-ai -space-x.

36 *Eliezer Yudkowsky:* Ibid.

36 *AI researcher Ben Goertzel:* Ben Goertzel, "Nine Years to a Positive Singularity," Singularity Summit, San Francisco, California, September 8, 2007, last accessed on July 31, 2017, http://web.archive .org/web/20130729203724id_/http://itc.conversationsnetwork.org /shows/detail3383.html.

38 Six Drivers of Global Change: Al Gore, *The Six Drivers of Global Change* (New York: Random House, 2013), 3–46.

38 *US Labor Department:* "Healthcare: Millions of Jobs Now and in the Future," *Occupational Outlook Quarterly*, US Department of Labor website, Spring 2014, last accessed on July 23, 2017, https:// www.bls.gov/careeroutlook/2014/spring/art03.pdf.

38 *displace around 3 million:* Olivia Solon, "Self-Driving Trucks: What's the Future for America's 3.5 Million Truckers," *Guardian*, June 17, 2016, last accessed on July 25, 2017, https://www.theguard ian.com/technology/2016/jun/17/self-driving-trucks-impact-on -drivers-jobs-us.

38 *PricewaterhouseCoopers (PWC) released:* "Will Robots Steal Our Jobs? The Potential Impact of Automation on the UK and Other Major Economies," *UK Economic Outlook*, March 2017, last accessed on July 2, 2017, http://www.pwc.co.uk/economic-services /ukeo/pwcukeo-section-4-automation-march-2017-v2.pdf.

38 *Buckminster Fuller argued:* "Fuller's Influence," Buckminster Fuller Institute, last accessed on July 25, 2017, https://www.bfi.org/about -fuller/biography/fullers-influence.

39 *dwindled to below 2 percent:* "Industry Employment and Output Projections to 2024," *Monthly Labor Review*, US Department of Labor, Bureau of Labor Statistics, December 2015, last accessed on July 23, 2017, https://www.bls.gov/emp/ep_table_201.htm.

40 *his nightmare scenarios:* Nick Bostrom, *Superintelligence: Paths, Dangers, Strangers* (Oxford: Oxford University Press, 2014), 130–64.

42 *Iran's development of nukes:* Robert Einhorn and Richard Nephew, "The Iran Nuclear Deal: Prelude to Proliferation in the Middle East," *Brookings*, May 31, 2016, last accessed on July 31, 2017, https://www.brookings.edu/research/the-iran-nuclear-deal-prelude -to-proliferation-in-the-middle-east/.

42 *partnered with China:* Adam Rawnsley, "Meet China's Killer Drones," *Foreign Policy*, January 14, 2016, last accessed on July 29, 2017, http://foreignpolicy.com/2016/01/14/meet-chinas-killer -drones/.

42 *placed the largest order on record:* Quwa Team, "Saudi Arabia Will License-Produce Chinese Armed Drones," *Quwa: Defence News & Analysis Group*, March 24, 2017, last accessed on July 31, 2017, http://quwa.org/2017/03/24/saudi-arabia-will-license-produce-chi nese-armed-drones/.

42 *developed the Anka:* ANKA Multi-Role ISR System, TAI: Turkish Aerospace Industries, January 25, 2013, last accessed on July 31, 2017, https://www.tai.com.tr/en/project/anka-medium-altitude-long -endurance-uav-system.

43 *US RQ-170 aerial vehicle:* David Cenciotti, "Iran Unveils New UCAV Modeled on Captured U.S. RQ-170 Stealth Drone," *Avi-*

ationist, October 2, 2016, last accessed on July 31, 2017, https://
theaviationist.com/2016/10/02/iran-unveils-new-ucav-modeled-on
-captured-u-s-rq-170-stealth-drone/.

43 *Prisoner's Dilemma:* "Prisoner's Dilemma," *Wikipedia*, last edited
on July 30, 2017, last accessed on August 1, 2017, https://en.wikipe
dia.org/wiki/Prisoner%27s_dilemma.

45 *safety and "explainability":* Several of my colleagues in the University of Texas's Robotics Department are focusing their work on
"explainability" and safety in autonomous control systems, including Professor Scott Niekum, Professor Ufuk Topcu, and Professor
Peter Stone, last accessed on July 29, 2017, https://robotics.utexas
.edu/people.

46 *program called xAI:* "Explainable Artificial Intelligence (xAI),"
DARPA (Defence Advanced Research Projects Agency), last
accessed on July 31, 2017, https://www.darpa.mil/program/explain
able-artificial-intelligence.

47 *James Barrat calls AI:* James Barrat, *Our Final Invention: Artificial
Intelligence and the End of the Human Era* (New York: Thomas
Dunne Books, 2013).

PART TWO: TODAY AND TOMORROW

1. THE EMERGING INTERNET OF THINGS

52 *inventor Joseph Marie Jacquard:* Michael N. Geselowitz, "The Jacquard Loom: A Driver of the Industrial Revolution," *The Institute:
The IEEE news source*, July 18, 2016, last accessed on July 31, 2017,
http://theinstitute.ieee.org/tech-history/technology-history/the
-jacquard-loom-a-driver-of-the-industrial-revolution.

2. HEALTHCARE

58 *participants begin to speak:* The following stories are based on
experiences shared during a support group meeting for sufferers of
Crohn's and colitis in Austin, Texas, in the fall of 2016. The participants requested to keep identities private.

59 *study of the gut:* NIH Human MicroBiome Project: DACC-Tools

and Protocols, last accessed on July 31, 2017, http://hmpdacc.org/resources/tools_protocols.php

60 Clostridium difficile, *or* C. diff: Lawrence J. Brandt, "Fecal Transplantation for the Treatment of *Clostridium difficile* Infection," *Gastroenterology & Hepatology* 8, no. 3 (2012): 191–94, last accessed on July 24, 2017, https://www.ncbi.nlm.nih.gov/pmc/articles/PMC3365524/.

60 *published in* Science Translational Medicine: Giada De Palma et al., "Transplantation of Fecal Microbiota from Patients with Irritable Bowel Syndrome Alters Gut Function and Behavior in Recipient Mice," *Science Translational Medicine* 9, no. 379 (2017), last accessed on July 25, 2017, http://stm.sciencemag.org/content/9/379/eaaf6397.

61 *spoke with Bill Colston:* Phone conversation with Bill Colston on Tuesday, November 29, 2016.

65 *October 28, 2016, Chinese:* David Cyranoski, "CRISPR Gene-Editing Tested in a Person for the First Time," *Nature*, November 15, 2016, last accessed on July 31, 2017, http://www.nature.com/news/crispr-gene-editing-tested-in-a-person-for-the-first-time-1.20988#auth-1.

65 *Society for Neuroscience:* R. Douglas Fields, "Wireless Brain Implant Allows 'Locked-In' Woman to Communicate," *Scientific American*, November 12, 2016, last accessed on July 26, 2017, https://www.scientificamerican.com/article/wireless-brain-implant-allows-ldquo-locked-in-rdquo-woman-to-communicate/.

3. SECURITY IN THE CYBER AGE

69 *Germany's famous V-2 team:* Paul Grigorieff, "The Mittelwerk/Mittelbau/Camp Dora/Mittelbau GmbH—Mittelbau KZ," V2Rocket: The A-4/V-2 Resource Site, last accessed on July 31, 2017, http://www.v2rocket.com/start/chapters/mittel.html.

70 *A number of investigations:* Ellen Nakashima, "Confidential Report Lists U.S. Weapons System Designs Compromised by Chinese Cyberspies," *Washington Post*, May 27, 2003, last accessed on July 31, 2017, https://www.washingtonpost.com/world/national-security/confidential-report-lists-us-weapons-system-designs-compromised-by-chinese-cyberspies/2013/05/27/a42c

3e1c-c2dd-11e2-8c3b-0b5e9247e8ca_story.html?utm_term=.c8d1
1fbfc6c4; Siobhan Gorman, August Cole, and Yochi Dreazen,
"Computer Spies Breach Fighter-Jet Project," *Wall Street Journal*,
April 21, 2009, last accessed on August 1, 2017, https://www.wsj
.com/articles/SB124027491029837401.

70 *Su Bin:* Ryan O'Hare, "China Proudly Debuts Its New Stealth Jet
Is Built 'by Hacking into US Computers and Stealing Plans,'" *Daily
Mail*, November 1, 2016, last accessed on July 23, 2017, http://www
.dailymail.co.uk/sciencetech/article-3893126/Chinese-J-20-stealth
-jet-based-military-plans-stolen-hackers-makes-public-debut.html
#ixzz4oZxNpW7e.

71 *Target and Home Depot's:* Kevin Granville, "Nine Recent Cyberattacks Against Big Business," *New York Times*, February 5, 2015,
last accessed on August 2, 2017, https://www.nytimes.com/interac
tive/2015/02/05/technology/recent-cyberattacks.html?_r=0.

71 *resignation of Iceland's prime minister:* Steven Erlanger, Stephen
Castle, and Rick Gladstone, "Iceland's Prime Minster Steps Down
Amid Panama Papers Scandal," *New York Times*, April 5, 2016, last
accessed on August 2, 2017, https://www.nytimes.com/2016/04/06
/world/europe/panama-papers-iceland.html.

71 *Turkish power grid:* Don Melvin, "Power Outage Hits Turkey,"
CNN, March 31, 2015, last accessed on August 1, 2017, http://www
.cnn.com/2015/03/31/middleeast/turkey-power-outage/index.html.

71 *Yahoo reported:* Robert McMillan, "Yahoo Says Information on at
Least 500 Million User Accounts Was Stolen," *Wall Street Journal*,
September 22, 2016, last accessed on July 31, 2017, https://www
.wsj.com/articles/yahoo-says-information-on-at-least-500-million
-user-accounts-is-stolen-1474569637.

71 *Charlie Miller and Chris Valasek revealed:* Andy Greenberg, "The
Jeep Hackers Are Back to Prove Car Hacking Can Get Much
Worse," *Wired*, August 1, 2016, last accessed on August 1, 2017,
https://www.wired.com/2016/08/jeep-hackers-return-high-speed
-steering-acceleration-hacks/.

72 *Gartner Research:* Gartner Press Release, "Gartner Says 8.4 Billion
Connected 'Things' Will Be in Use in 2017, Up 31 Percent from
2016," *Gartner Research*, February 7, 2017, last accessed on August
2, 2017, http://www.gartner.com/newsroom/id/3598917.

72 *Barack Obama:* Barack Obama, "Remarks by the President at the

National Cybersecurity Communications Integration Center," National Cybersecurity Communications Integration Center, Arlington, Virginia, January 13, 2015, last accessed on August 2, 2017, https://obamawhitehouse.archives.gov/the-press-office/2015/01/13 /remarks-president-national-cybersecurity-communications-inte gration-cent.

72 *$19 billion:* Dustin Volz, "U.S. Government Worse Than All Major Industries on Cyber Security: Report," Reuters, April 14, 2016, last accessed on July 31, 2017, http://www.reuters.com/article/us-usa -cybersecurity-rankings-idUSKCN0XB27K.

73 *Office of Personnel Management:* Julie Hirschfeld Davis, "Hacking of Government Computers Exposed 21.5 Million People," *New York Times*, July 9, 2015, last accessed on August 1, 2017, https:// www.nytimes.com/2015/07/10/us/office-of-personnel-manage ment-hackers-got-data-of-millions.html.

73 *Bill Mellon:* Richard Perez-Pena, "Universities Face a Rising Barrage of Cyberattacks," *New York Times*, July 16, 2013, last accessed on July 31, 2017, http://www.nytimes.com/2013/07/17/education /barrage-of-cyberattacks-challenges-campus-culture.html.

74 *Elk Cloner:* Yoni Heisler, "World's first computer virus hit the Apple II 35 years ago," *BGR*, March 20, 2017, last accessed on August 1, 2017, http://bgr.com/2017/03/20/malware-first-computer-virus-apple -elk-cloner/.

75 *Kaspersky Lab:* "Kaspersky Lab: 323,000 New Malware Samples Found Each Day," *DARKReading*, December 7, 2016, last accessed on August 1, 2017, http://www.darkreading.com/vulnerabilities ---threats/kaspersky-lab-323000-new-malware-samples-found-each -day/d/d-id/1327655.

75 *According to Mandiant:* Nicole Perlroth, "Hackers in China Attacked The Times for Last 4 Months," *New York Times*, January 30, 2013, last accessed on August 2, 2017, http://www.nytimes.com/2013/01/31 /technology/chinese-hackers-infiltrate-new-york-times-computers .html.

76 *Cyber Grand Challenge:* "The World's First All-Machine Hacking Tournament," DARPA, August 4, 2016, last accessed on August 1, 2017, http://archive.darpa.mil/cybergrandchallenge/.

76 *massive ransomware attack:* Lily Hay Newman, "The Ransomware Meltdown Experts Warned About Is Here," *Wired*, May 12, 2017,

last accessed on July 26, 2017, https://www.wired.com/2017/05/ran somware-meltdown-experts-warned/.

76 *WannaCry:* Bryan Lares, "Ransomware Attacks Healthcare: Deep Armor Catches the Culprit, 'WannaCry' Malware," *SparkCognition Blog,* May 12, 2017, last accessed on August 28, 2017, https:// sparkcognition.com/2017/05/ransomware-attacks-healthcare-deep armor-catches-culprit-wannacry-malware/.

78 *healthcare is now one of the industries:* Lily Hay Newman, "Medical Devices Are the Next Security Nightmare," *Wired,* March 2, 2017, last accessed on July 31, 2017, https://www.wired.com/2017/03 /medical-devices-next-security-nightmare/.

82 Like Sherlock Holmes*:* Maria Konnikova, *How to Think Like Sherlock Holmes* (New York: Penguin, 2013), 239.

84 *Adylkuzz:* Brian Lares, "Adylkuzz, the Latest Zero-Day Threat Malware, Detected by DeepArmor Enterprise," *SparkCognition Blog,* May 24, 2017, last accessed on August 28, 2017, https://sparkcogni tion.com/2017/05/adylkuzz-malware-deeparmor/.

84 *"EternalBlue" exploit:* Kafeine, "Adylkuzz Cryptocurrency Mining Malware Spreading for Weeks Via EternalBlue/DoublePulsar," *proofpoint,* May 15, 2017, last accessed on August 2, 2017, https:// www.proofpoint.com/us/threat-insight/post/adylkuzz-cryptocur rency-mining-malware-spreading-for-weeks-via-eternalblue-double pulsar.

85 *Third US Offset Strategy:* Deputy Secretary of Defense Bob Work, "The Third U.S. Offset Strategy and Its Implications for Partners and Allies," Willard Hotel, Washington, D.C., January 28, 2015, last accessed on August 1, 2017, https://www.defense.gov/News /Speeches/Article/606641/.

86 *a hundred small drones:* Chris Baraniuk, "US Military Tests Swarm of Mini-Drones Launched from Jets," BBC, January 10, 2017, last accessed on August 2, 2017, http://www.bbc.com/news/technology -38569027.

4. WARFARE AND AI

87 *General John Allen:* General John R. Allen and Amir Husain, "On Hyperwar," *Proceedings Magazine* (U.S. Naval Institute) 143, no. 7 (July 2017): 1,373.

90 *Psychologist Daniel Kahneman:* Daniel Kahneman, *Thinking Fast and Slow* (New York: Farrar, Straus and Giroux, 2013), 43.

102 *hydrofoil boats:* Jon Stock, "Little Boat, Big Danger: How a British-Made Speedboat Has Become a Weapon in Iran's Standoff with the US," *Telegraph*, August 20, 2012, last accessed on July 31, 2017, http://www.telegraph.co.uk/news/worldnews/middleeast /iran/9486815/Little-boat-big-danger-how-a-British-made-speed boat-has-become-a-weapon-in-Irans-standoff-with-the-US.html.

102 *state-run* China Daily: Zhao Lei, "Nation's Next Generation of Missiles to Be Highly Flexible," *China Daily*, August 18, 2016, last accessed on August 1, 2017, http://www.chinadaily.com.cn /china/2016-08/19/content_26530461.htm.

103 *Ken Minihan:* Phone conversation with Ken Minihan on September 2, 2015.

104 *Chinese government orchestrated:* Robert Windrem, "Secret NSA Map Shows China Cyber Attacks on U.S. Targets," NBC News, July 30, 2015, last accessed on August 2, 2017, http://www.nbcnews .com/news/us-news/exclusive-secret-nsa-map-shows-china-cyber -attacks-us-targets-n401211.

104 The Science of Military Strategy: Shane Harris, "China Reveals Its Cyberwar Secrets," *Daily Beast*, March 18, 2015, last accessed on August 2, 2017, http://www.thedailybeast.com/china-reveals-its-cyber war-secrets.

105 *largest next-generation sequencing:* John Fox and Jim King, "Chinese Institute Makes Bold Sequencing Play," *Nature Biotechnology* 28 (2010): 189–91, last accessed on July 31, 2017, http://www .nature.com/nbt/journal/v28/n3/full/nbt0310-189c.html?foxtrotcall back=true.

105 *use of "backscatter":* Bart Jansen, "TSA Defends Full-Body Scanners at Airport Checkpoints," *USA Today*, March 2, 2016, last accessed on August 2, 2017, https://www.usatoday.com/story/news /2016/03/02/tsa-defends-full-body-scanners-airport-checkpoints /81203030/.

5. FINANCIAL MARKETS

109 *John Maynard Keynes:* Colin L. Read, "For John Maynard Keynes, Economic Theory Was a Sideline," *BloombergView*, November 2,

2012, last accessed on August 2, 2017, https://www.bloomberg.com /view/articles/2012-11-02/for-john-maynard-keynes-economic -theory-was-a-sideline.

110 *"blood in the streets":* Daniel Myers, "Buy When There's Blood in the Streets," *Investopedia*, March 7, 2017, last accessed on August 2, 2017, http://www.investopedia.com/articles/financial-theory/08 /contrarian-investing.asp.

115 Wired *magazine:* Cade Metz, "The Sadness and Beauty of Watching Google's AI Play Go," *Wired*, March 11, 2016, last accessed on August 2, 2017, https://www.wired.com/2016/03/sadness-beauty -watching-googles-ai-play-go/.

117 *discretionary hedge funds:* John Authers and Mary Child, "Hedge Funds: Overpriced, Underperforming," *Financial Times*, May 24, 2016, last accessed on August 2, 2017, https://www.ft.com/content /9bd1150e-1b76-11e6-b286-cddde55ca122; Adam Sarhan, "Is Discretionary Macro Dead?," *Forbes*, August 17, 2016, last accessed on August 2, 2017, https://www.forbes.com/sites/adamsarhan/2016 /08/17/is-discretionary-macro-dead-wheres-joe-dimaggio/#3bffb 1935120; Suzanne McGee, "Rise of the Billionaire Robots: How Algorithms Have Redefined Hedge Funds," *Guardian*, May 15, 2016, last accessed on July 31, 2017, https://www.theguardian.com /business/us-money-blog/2016/may/15/hedge-fund-managers-algo rithms-robots-investment-tips.

117 *2016 Goldman Sachs study:* Rachael Levy, "This Is the Biggest Trend in the Hedge Fund World Right Now," *Business Insider*, August 17, 2016, last accessed on July 31, 2017, http://www.business insider.com/quant-investing-is-the-biggest-new-trend-for-hedge -funds-2016-8.

120 *reporter Leah McGrath Goodman:* Leah McGrath Goodman, "The Face Behind Bitcoin," *Newsweek*, March 6, 2014, last accessed on July 31, 2017, http://www.newsweek.com/2014/03/14/face-behind -bitcoin-247957.html.

120 *Craig Steven Wright:* Andy Greenberg, "How Craig Wright Privately Proved He Created Bitcoin," *Wired*, May 2, 2016, last accessed on July 31, 2017, https://www.wired.com/2016/05/craig -wright-privately-proved-hes-bitcoins-creator/.

6. COGNITIVE SPACES

130 *America's Future Energy Jobs:* National Commission on Energy Policy's Task Force on America's Future Energy Jobs, 2009, Bipartisan Policy Center, last accessed on July 31, 2017, http://bipartisanpolicy.org/wp-content/uploads/sites/default/files/NCEP%20Task%20Force%20on%20America%27s%20Future%20Energy%20Jobs%20-%20Final%20Report.pdf.

133 *Synthetic Sensors:* Elizabeth Stinson, "A Sensor That Could Make Homes Scary Smart," *Wired*, May 11, 2017, last accessed on August 2, 2017, https://www.wired.com/2017/05/supercharged-sensor-soon-make-homes-scary-smart/.

133 *Nest in 2014:* Lance Whitney, "Google Closes $3.2 Billion Purchase of Nest," *CNET*, February 12, 2014, last accessed on July 31, 2017, https://www.cnet.com/news/google-closes-3-2-billion-purchase-of-nest/.

133 *acquisition of Dropcam:* Marco Chiappetta, "Google Nest Lab's Acquisition of Dropcam Scares the Heck Out of Me," *Forbes*, June 23, 2014, last accessed on July 31, 2017, https://www.forbes.com/sites/marcochiappetta/2014/06/23/google-nest-labs-acquisition-of-dropcam-scares-the-heck-out-of-me/#44691ecb5904.

134 *range of systems:* Hope Reese, "CES 2017: Robots of the Future," *TechRepublic*, January 6, 2017, last accessed on July 31, 2017, http://www.techrepublic.com/article/ces-2017-robots-of-the-future/.

7. MIND HACKING

144 *Rayid Ghani:* Gregory Piatetsky, "KDnuggets Exclusive: Interview with Rayid Ghani, Chief Scientist Obama 2012 Campaign," *KDnuggets*, January 2013, last accessed on July 31, 2017, http://www.kdnuggets.com/2013/01/kdnuggets-exclusive-interview-rayid-ghani-chief-scientist-obama-2012-campaign.html.

144 *mass psychographic profiling:* Aditya Madanapalle, "Big Data and Psychographic Profiling Helped Donald Trump Win the US Presidential Election," *Firstpost*, January 31, 2017, last accessed on July 29, 2017, http://www.firstpost.com/tech/news-analysis/big-data-and-psychographic-profiling-helped-donald-trump-win-the-us-presidential-election-3696847.html.

145 *article in* Motherboard*:* Hannes Grassegger and Mikael Kro-
gerus, "The Data That Turned the World Upside Down," *Motherboard*, January 28, 2017, last accessed on August 2, 2017, https://
motherboard.vice.com/en_us/article/mg9vvn/how-our-likes-helped
-trump-win.

145 *to widespread recognition:* Nicholas Confessore and Danny Hakim,
"Data Firm Says 'Secret Sauce' Aided Trump; Many Scoff," *New
York Times*, March 6, 2017, last accessed on August 2, 2017, https://
www.nytimes.com/2017/03/06/us/politics/cambridge-analytica
.html?_r=0; Matea Gold and Frances Stead Sellars, "After Working
for Trump's Campaign, British Data Firm Eyes New U.S. Government Contracts," *Washington Post*, February 17, 2017, last accessed
on July 31, 2017, https://www.washingtonpost.com/politics/after
-working-for-trumps-campaign-british-data-firm-eyes-new-us
-government-contracts/2017/02/17/a6dee3c6-f40c-11e6-8d72
-263470bf0401_story.html?utm_term=.c630f34038fe.

145 *where Jared Kushner:* Steven Bertoni, "Exclusive Interview: How
Jared Kushner Won Trump the White House," *Forbes*, December 20, 2016, last accessed on August 1, 2017, https://www.forbes
.com/sites/stevenbertoni/2016/11/22/exclusive-interview-how-jared
-kushner-won-trump-the-white-house/#78a4f0893af6.

146 ScienceNode*:* Tristan Fitzpatrick, "Data Wins the Day: How HPC
Turned the Tide for Trump," *ScienceNode*, November 17, 2016, last
accessed on August 2, 2017, https://sciencenode.org/feature/how
-polling-models-failed-to-accurately-predict-the-outcome-of-the
-election.php.

146 *voter suppression ads:* Joshua Green and Sasha Issenberg, "Inside
the Trump Bunker with Days to Go," *Bloomberg Businessweek*,
October 27, 2016, last accessed on July 31, 2017, https://www
.bloomberg.com/news/articles/2016-10-27/inside-the-trump-bun
ker-with-12-days-to-go; Joel Winston, "How the Trump Campaign Built an Identity Database and Used Facebook Ads to Win
the Election," *StartupGrind*, November 18, 2016, last accessed on
August 2, 2017, https://medium.com/startup-grind/how-the-trump
-campaign-built-an-identity-database-and-used-facebook-ads-to
-win-the-election-4ff7d24269ac.

147 *$191 billion in 2016:* Lucy Handley, "Global Advertising Spend to
Slow in 2017; While 2016 Sales Reached Nearly $500bn: Research,"
CNBC, December 5, 2016, last accessed on August 2, 2017, https://

www.cnbc.com/2016/12/05/global-ad-spend-to-slow-in-2017-while
-2016-sales-were-nearly-500bn.html.

148 *forecasters predict:* "Digital Advertising Revenues to Double by
2020, Rising to $285 Billion," *Jupiter Research*, June 20, 2016, last
accessed on August 2, 2017, https://www.juniperresearch.com/press
/press-releases/digital-advertising-revenues-to-double-by-2020.

149 *CNN reported:* Ivana Kottasova, "A Third of Pro-Trump Tweets Are
Generated by Bots," *CNN Tech*, October 18, 2016, last accessed on
August 1, 2017, http://money.cnn.com/2016/10/18/technology/twit
ter-bots-donald-trump-hillary-clinton/index.html.

150 American Journal of Public Health *published:* David D. Luxton,
Jennifer D. June, and Jonathan M. Fairall, "Social Media and Suicide: A Public Health Perspective," *American Journal of Public
Health* 102 (May 2012): S195–S200, last accessed on July 31, 2017,
https://www.ncbi.nlm.nih.gov/pmc/articles/PMC3477910/.

152 *"the backfire effect":* Brendan Nyhan, Jason Reifler, Sean Richey,
and Gary L. Freed, "Effective Messages in Vaccine Promotion: A
Randomized Trial," *Pediatrics*, March 3, 2014; Brendan Nyhan,
Jason Reifler, and Peter Ubel, "The Hazards of Correcting Myths
about Health Care Reform," *Medical Care* 51, no. 2 (2013): 127–32.

152 *Daniel Kahneman:* Daniel Kahneman, *Thinking, Fast and Slow*
(New York: Farrar, Straus and Giroux, 2013).

152 *"anchoring effect":* Amos Tversky and Daniel Kahneman, "Judg-
ment under Uncertainty: Heuristics and Biases," *Science* 185, no.
4157 (1974): 1124–31.

154 *Robert Epstein:* Robert Epstein and Ronald E. Robertson, "The
Search Engine Manipulation Effect (SEME) and Its Possible
Impact on the Outcomes of Elections," *PNAS* 112, no. 33 (2015):
E4512–21.

155 *Peter Corning:* Peter Corning, "The Re-Emergence of 'Emergence': A Venerable Concept in Search of a Theory," *Complexity* 7,
no. 6 (2002): 18–30.

NOTES

PART THREE: THE FUTURE

8. THE MISSING BLOCK . . .

166 *"The Last Question":* Isaac Asimov, *The Complete Stories,* Vol. 1 (New York: Doubleday, 1990), 290–300.

9. DECOUPLING WORK AND PURPOSE

169 *historian Yuval Harari:* Yuval Harari, *Sapiens: A Brief History of Humankind* (New York: HarperCollins, 2015).

170 *Oxfam reported:* "An Economy for the 1%," *Oxfam,* January 18, 2016, last accessed on August 2, 2017, https://www.oxfam.org/en /research/economy-1.

171 *universal basic income:* Sonia Sodha, "Is Finland's Basic Universal Income a Solution to Automation, Fewer Jobs and Lower Wages?," *Guardian,* February 19, 2017, last accessed on July 31, 2017, https:// www.theguardian.com/society/2017/feb/19/basic-income-finland -low-wages-fewer-jobs.

10. THE PURSUIT OF KNOWLEDGE

174 *"I was a Treasure":* 1000 *Qudsi Hadiths: An Encyclopedia of Divine Sayings* (New York: Arabic Virtual Translation Center, 2012).

11. GENESIS AI

180 *"perceptronium":* Max Tegmark, "Consciousness as a State of Matter," *Chaos, Solitons & Fractals* 76 (2015): 238–70.

FURTHER READING

Barrat, James. *Our Final Invention: Artificial Intelligence and the End of the Human Era.* New York: Thomas Dunne Books, 2013.

Bostrom, Nick. *Superintelligence: Paths, Dangers, Strangers.* Oxford: Oxford University Press, 2014.

Christian, Brian, and Tom Griffiths. *Algorithms to Live By: The Computer Science of Human Decisions.* New York: Henry Holt, 2016.

Domingos, Pedro. *The Master Algorithm: How the Quest for the Ultimate Learning Machine Will Remake Our World.* New York: Basic Books, 2015.

Harari, Yuval. *Sapiens: A Brief History of Humankind.* New York: HarperCollins, 2015.

Kurzweil, Ray. *The Age of Spiritual Machines.* New York: Penguin, 2000.

Minsky, Marvin. *The Society of Mind.* New York: Simon & Schuster, 1988.

Penrose, Roger. *The Emperor's New Mind: Concerning Computers, Minds and the Laws of Physics.* Oxford: Oxford University Press, 2016.

Rid, Thomas. *Rise of the Machines: A Cybernetic History.* New York: W. W. Norton, 2016.

Tegmark, Max. *Our Mathematical Universe: My Quest for the Ultimate Nature of Reality.* New York: Knopf, 2014.

INDEX

INDEX

ABOUT THE AUTHOR

Amir Husain is an award-winning serial entrepreneur and inventor based in Austin, Texas. He is the founder and CEO of SparkCognition, Inc., a company specializing in artificial intelligence platforms that help businesses and governments better respond to a world of ever-evolving threats. He was a founding advisory board member for IBM's Watson Cognitive Computing platform and speaks at numerous SXSW, defense, cybersecurity, computer science, energy, and environmental conferences. Amir and SparkCognition's work has been featured in publications such as *Fast Company*, *Wired*, *Forbes*, and the *New York Times*. *The Sentient Machine* is his first book.